Silence within and beyond Pedagogical Settings

Eva Alerby

Silence within and beyond Pedagogical Settings

palgrave
macmillan

Eva Alerby
Department of Arts, Communication and Education
Luleå University of Technology
Luleå, Sweden

ISBN 978-3-030-51059-6 ISBN 978-3-030-51060-2 (eBook)
https://doi.org/10.1007/978-3-030-51060-2

This Palgrave Pivot imprint is published by the registered company Springer Nature Switzerland AG.
The registered company address is: Gewerbestrasse 11, 6330 Cham, Switzerland

Originally published 2012 by the publishing house Studentlitteratur,
Lund, Sweden.
Swedish title: *Om tystnad – i pedagogiska sammanhang.*
Translation by Edward Caplen

FOREWORD

Educational institutions these days are dominated by texts, and yet, even though reading and writing seem such quiet activities, they are still such *noisy* places. Eva Alerby has been working for many years to tease out the value of silence, the contribution of silence to educational endeavours and the danger of seeing silence as merely a form of control and oppression. This fine book brings together her thoughts on what silence is, how it is used and why it should be appreciated.

Finding the opportunities for silence is an important task for schools. Extended 'silent reading' is not as popular as it once was, and yet 'silent discos' have recently been introduced to festivals and clubs, where people wear headphones to listen to music, and dance together. Such silent discos are only silent to observers: they are filled with music for each participant. The same might be said of 'silent reading' in schools: an observer hears little, but each reader's head is filled with the voices of the books being read. So perhaps all the reading and writing that dominates education is indeed the source of a lot of noise—the barely subdued voices of authors with plenty to say.

And the people 'voiced' in texts also provide us with silences. As Eva Alerby says, the gaps between the words make a text readable. Poetry, with its shortened lines, is—as T. S. Eliot said—'writing with a lot of silence on the page.' Music, too, depends on silence as much as it does on sounds.

If noise can be found in apparently quiet activities like reading, and silence can be found in the midst of noise, what does this tell us? Silence is hard to pin down, certainly. And—counter-intuitively—it requires the ability to hear and to listen carefully. Silence is only ever experienced alongside noise. But educational institutions such as schools too often try to block out the natural healthy silences that make noise so interesting. Teachers may ask for instant answers, and may even think that their own voices are the only ones that matter: they are not providing opportunities for healthy silence. They are creating silences of meek obedience. Even seeing themselves as 'speakers' can be problematic for teachers. The best teachers are in dialogue, and as such are able to listen attentively to what their students have to say, and thereby 'listen their students into expression.' As Martin Buber, the philosopher of dialogue, says, a teacher engaged in 'mere instruction,' rather than a 'real lesson,' is not engaged in dialogue, but 'monologue disguised as dialogue.'

Eva Alerby has written a book that quietly asserts, and voices, the value of silence. The author exhibits a calm assurance that also—surreptitiously, quietly—provokes and politicises. This book enables deep thought on an important dimension of education, on the bland silence described merely as the absence of sound, on the oppressive silence that 'unvoices' people and, most of all, on the quietude of peaceful contemplation that prepares us for action. Quietude is not an alternative to making the world a better place: it is a necessary dimension *of* action. Sports coaches and music tutors alike describe the need to 'get in the zone' in preparing for a competition or performance. Many centuries ago, the *Bhagavad-Gītā* described the need for quiet withdrawal, as a tortoise withdraws its head and legs into its shell, in order to attain the right ('enstatic') frame of mind to go into battle. So I see this book as a philosopher's book, a teacher's book and an activist's book, preparing its readers to be sufficiently quiet to hear what is needed—from within ourselves and from those whose voices are too often suppressed.

York, UK
February 2020   Julian Stern

A Word from the Author: A Walk Along the Shore of the River Loire

Early one morning during a visit to France, I was walking alone along the shore of the river Loire. The gentle flow of the water and the peace and quiet gave me a much-needed sense of calm. Isn't it wonderful to just walk in silence? But, of course, the question is whether it really *was* completely quiet—what do we really mean when we say that it is quiet, or that we have a longing for silence? These are questions that will be illuminated and considered later on in this publication.

In stark contrast to this enjoyable experience of silence is an event that took place some years ago as I was travelling through another country—a county whose language I had not mastered sufficiently in order to converse at any reasonably advanced level. In spite of this, I found myself in just such a situation where I was expected to take part in a discussion. I could understand just enough to be able to get the general gist of the conversation, and I even had opinions on the issue at hand, but I lacked the language skills needed in order to express myself in the way I wanted. As my brain switched into high gear, I could hear within myself what I wanted to say, but instead of my thoughts turning into expressions, I became silent. In the beginning, nobody noticed my silence, but after a while, people began to make comments about it until, in the end, some of them started to talk about and question my reticence. The situation became more and more uncomfortable, and I felt incredibly frustrated at not being able to express what I really wanted to say. That time, silence was both forced and uncomfortable.

It was this event that got me thinking about silence, and the result was that, over the years, I have presented and discussed a number of papers at conferences, published articles, book chapters as well as a book on silence—always with a connection to various pedagogical settings. But let us take a moment now to stop and think about the seeming contradiction of talking (or writing) about silence. And even though Martin Heidegger believed that the most objectionable and intolerable babble is produced when people discuss or write about silence, that is exactly what I shall now attempt to do. But allow me to return to this in the conclusion of the book.

There is, of course, a great variety of dimensions when it comes to silence within and beyond different pedagogical settings that could be expressed and written in a book such as this. But sometimes it is, as Max van Manen puts it, better to leave things unsaid; or in this case unwritten, instead of spelling everything out in black and white—to allow the text to be quiet. The silent spaces between the text are sometimes just as important as the words surrounding them. There are also those who believe that the one who does not have anything to say or write easily becomes verbose.

The book you are holding was originally published in Swedish by Studentlitteratur in 2012. Now that it is being reprinted in English, the text has been slightly revised and brought up to date, and the paragraph about silence and solitude in the light of texture is completely new. Some parts of the book originate from papers and articles that were produced in collaboration with others, and the text has been adapted to match the content of this book. So, I would like to extend a warm and sincere thanks to my former co-authors Jórunn Elídóttir and Catrine Kostenius for such good and creative collaboration and for your generosity in allowing me to 'recycle' parts of our previous work. I would also like to express my gratitude to all my other friends and colleagues around the world who contributed with constructive suggestions for improvement during the creation of this book—many thanks! Publishing house Palgrave Macmillan also deserves my appreciation for believing in this book project. Last but not least, I extend a warm and heartfelt thanks to Professor Julian Stern from York St John University in the UK, who took the time and effort to write the Foreword to the book. My genuine hope is that this text will contribute to contemplation and reflection on silence.

Villvattnet (2937 km from the shore of the river Loire)—one evening in February 2020, when the snow glittered in the dancing light of Aurora Borealis set to the backdrop of an ink-black sky, as the thermometer showed −29 °C, and all was quiet.

Eva Alerby

CONTENTS

Prologue: '... to be able to touch souls'

'See how nature—trees, flowers, grass—grows in silence; see the star, the moon and the sun, how they move in silence … we need silence to be able to touch souls,' says Mother Teresa (Bell & Battin, 1995, p. 104). To have the capacity to touch another person's soul is probably one of the most essential abilities people need when they meet and interact with one another—that is, if it should be a true and genuine meeting between them. Although there are, of course, exceptions, for the most part, we humans meet many others every day that we only see in passing. We meet at the supermarket till, on the bus, when we use a pedestrian crossing, at the fuel station, when queuing for the cash machine or in the dentist's waiting room. And if we are in school, either as a student or as a teacher, we meet many others during the school day. We might not even see or notice many of these people, whilst we give others a silent look that can signal empathy and understanding, or perhaps irritation or fear. With other people, though, we have another sort of relationship; a relationship where we truly meet and where we touch each other's souls, to borrow the words of Mother Teresa. Hopefully, this is something that happens in our schools; that the souls of students—and even the teachers, for that matter—are touched in some way. We could, perhaps, go so far as to say that 'to be able to touch souls' is the foundation of genuine human relations, and, according to Mother Teresa, silence is needed in order to achieve that. So, what do we know about silence? How well does the proverb 'speech is silver, but silence is golden' fit with today's society in general, and—more specifically—in our schools? Is it true that silence is more highly valued than the spoken word at school? What do we know about the role of silence in different pedagogical settings, and does silence always show the glimmer of gold? But allow me, please, to return to these questions later and instead begin by reflecting on whether we need to have insight regarding silence and its role for us humans, and, if so, why it is needed.

In one form or another, silence is always around us. The question is only whether we notice it. Sometimes, silence can be deafening and impossible to avoid, whilst at other times, it is not even noticed. In order to hear the silence, we may perhaps need to listen to it. Sometimes, silence is something longed for and desirable, and sometimes it is a mere nuisance or discomforting. Different people can perceive the same silent situation in completely different ways. Von Wright (2012, p. 94), however, emphasises that '[s]ilence "as such" is neither good nor bad.'

Regardless of whether silence is perceived as being good or bad, as pleasant or unpleasant, it must be regarded as an abstract phenomenon. Silence can be neither seen nor touched, but it is often described as something tangible, something we can see, hear and touch. For example: 'a wall of silence,' 'silent as a clam,' 'tumultuous silence,' 'the veil of silence' and so on. Some other expressions we use are 'a silent minute,' 'silent revolution' and 'deep silence.' These phrases describe silence as an emotive phenomenon, yet in a metaphoric form.

Another question I would like to raise is whether it matters who it is that is quiet and why they are. Sometimes, a person can be quiet of their own volition, whilst, for others, silence can be imposed upon them. Someone may be silent in order to demonstrate their position of power or superiority in relation to others, but someone else could be quiet because of not having anything to say or add to a conversation. Wittgenstein (1922) has the following to say in connection with this subject: 'What can be said at all can be said clearly; and whereof one cannot speak, thereof one must be silent' (p. 23).

Something to reflect upon in this context is who it is that sets the rules for what kind of silence is, or is not, acceptable. What values and societal structures provide acceptance for silence, and which ones do not? To understand and have insight in the meaning of silence and its various aspects is of great importance, not only in different pedagogical settings but for life itself. Therefore, whilst this book will discuss the importance of silence mostly in various pedagogical settings, the subject will also be explained and discussed from a more general perspective.

From the above, it can be seen that the main theme of this book is *silence*. Silence will be treated as both an individual phenomenon and in connection with various pedagogical settings, which is the reasoning from which the title of this book emerges: *Silence Within and Beyond Pedagogical Settings.*

PRESENTATION OF WHAT IS TO COME

This book is divided into three parts. Part I 'Silence in the Light of Theoretical Reflections' consists of four chapters, starting with a brief, general overview of some of the different aspects of silence. Throughout the text, I will argue that silence conveys a message—that it is important to listen to silence and at the same time to be open and responsive to its message. Another aspect that will be highlighted is one of the contradictions of silence: sound. Still other aspects I will consider are how silence can be explained from a linguistic and communicative perspective. The relationship between silence and power, silence in the light of cultural perspectives and voluntary silence, or the desire to be in silence, will then be outlined and discussed, followed by a review of the multifaceted nature of silence in the form of literal, epistemological and ontological silence.

Some alternatives to the spoken word that take the form of various arts will be examined more closely. Specifically, I will feature some silent, or nonverbal, modes of expression: the written word, including poetry; textures; images; music; and pantomime and mime. The intention is to reflect on and problematise these modes of expression in relation to both silence and educational situations.

The phenomenon of listening, along with the immanent importance of silence in relation to it, will be explored. For people in general, and for teachers in particular, it is important to really listen to students, and silence is a crucial aspect of listening. Listening, in turn, is an aspect that is valued in school settings.

How silence can be viewed in schools and other pedagogical settings will also be elucidated in more detail; for example, the way silence can be used, amongst other things, as an educational strategy, in which the value of allowing silent breaks and opportunities for both giving and taking silence is an essential part.

Part II of the book, 'Considerations of Silence in Day-to-Day Life at School: Some Experiences from Students,' which consists of two chapters, highlights experiences that students have had regarding the importance of silence in school. Various forms of silence and/or silenced messages are then discussed and exemplified with the help of school students who responded to a survey.

In the concluding part (Part III), 'The Art of Appreciating Silence,' previous reasoning is rounded off with a chapter highlighting some final reflections on silence and pedagogy: How the view of silence can change over time; how silence is accepted in today's society in general, and especially in today's schools; and also whether nonverbal expressions can speak or whether they should be considered silent. Further aspects I cover are whether, and in that case, how, silence is valued in school, and also what is said in what is not said. Finally, I reflect on the question: Is it at all possible to speak (or write) about silence?

LIST OF FIGURES

PART I

Silence in the Light of Theoretical Reflections

CHAPTER 1

Speaking Silence: The Different Aspects of Silence

'When they remain silent, they cry out' is a translation of the Latin expression *cum tacent clamant*. An equivalent expression that also captures this contradiction is 'speaking silence.' The Latin expression was used to describe the senators' reaction to Cicero's orations against Catiline in 62 BC—their silence speaks louder than words (EUdict dictionary, 2019). Can silence be eloquent? If so, what is silence saying to us? Is it possible for a person to say something without speaking? What are students saying when they stay quiet instead of answering a teacher's question?

Many thinking people have reflected on the idea of speaking and being silent. True dialogue can be both spoken and mute, and the latter is actually sometimes referred to as speaking silence (Buber, 1993). A person can speak continuously without actually saying anything, but another might say a great deal without speaking a word or expressing himself verbally in any way (Heidegger, 1971). Not saying anything—to stay quiet—is sometimes what says the most (Dickinson, 2020), and it is also possible to say that even a 'non-message' is a message (Bateson, 1987).

Quite simply, silence seems to tell us something, as can be seen in the case of two acquaintances who meet each other on the street and one does not return the other's greeting, or when a student meets some classmates in the school corridor and greets them, but does not receive a greeting in return. Another example is when someone sends a letter that never receives a reply. When silence is preferred or forced, the interpretation and

© The Author(s) 2020
E. Alerby, *Silence within and beyond Pedagogical Settings*,
https://doi.org/10.1007/978-3-030-51060-2_1

conclusion could be that silence becomes a language when spoken words do not suffice. So how can silence be viewed?

In this chapter, I will highlight some of the various aspects of silence. I will argue that silence conveys something. But, in order to understand it, we must listen to the silence and be open and responsive to its message. One of the aspects I intend to highlight is an opposite of silence—sound. The other aspects are how silence can be explained from a linguistic and communicative perspective. The various characteristics of silence will also be discussed—is the silence perceived as good or bad, as friendly or malicious? I will then consider the relationship between silence and power, silence in the light of cultural perspectives and self-imposed silence or the desire to find it. The chapter concludes with a review of the multifaceted nature of silence in the form of literal, epistemological and ontological silences.

THE CONNECTION BETWEEN SILENCE AND SOUND

Normally, silence is considered to be the absence of sound—that is, the lack of auditory stimulation, a state of soundlessness. Is there, though, ever really a complete lack of sound? A total, absolute silence is contradicted by the fact that the human body always produces sound, such as through blood circulation, breathing and digestion, and as long as a person lives, these sounds are carried with them (Arlinger, 1995). But according to The National Encyclopedia (2010a), silence is described as 'a state of (almost) complete absence of sound.'

In order to gain an understanding of the meaning of silence, it may be useful to briefly consider one of the opposites of silence: sound. Sound is described by The National Encyclopedia (2010b) as fluctuations of pressure that propagate in the form of waves. In the narrow, original sense, this is limited to air fluctuations at a frequency of between 20 and 20,000 hertz. It is only sound that exists within this range that can be perceived by the human ear (Arlinger, 1995). Some say that sound levels below 20 decibels can be considered silent. Others argue that sound levels in a range between 10 and 20 decibels can be considered silent, even though the faintest sound a person with good hearing can detect is actually 0 decibels (Boverket, 2011; Hellberg, 2002). What is considered silent or not is more of a subjective experience. In a study conducted by Bergmark and Kostenius (2012), students described classroom noise level, and they felt that, whilst silence is important for fostering a positive learning climate, there is no need for complete silence.

If we use the description of silence as being the absence of sound, we might add that some noises die out—they become quiet over time. In spite of this, some of these extinct sounds live on. Some examples of sounds that have now become quiet are the clatter of a typewriter, the 'clickety-clack' of train wheels as they roll over joints in the rails, the sound a TV makes when it is switched off and the noise of a camera's shutter. These sounds have disappeared as technology has improved. The clattering keys of a typewriter have been replaced by the computer keyboard, joints in train rails have been minimised and almost completely removed, the technology used in televisions has advanced and the old mechanical on/off switches that used to produce the characteristic sound of a TV turning on or off have today been replaced with digital touch-sensitive controls that do not make sound in the same way.

Some of these classic sounds, though, have a tendency to stick around; like the shutter sound we hear when taking a picture with a digital camera. Today, few consumer digital cameras have shutters, but they can still sound like the old cameras. The difference is that the modern cameras we use today produce the sound digitally, it is a recorded sound stored on a chip inside the camera. This is because we expect a camera to sound a certain way—it shouldn't be completely silent.

On the subject of hertz and decibels, we need to stop and think about their meaning in relation to silence in pedagogical settings. We regularly hear alarming reports from preschools and schools about how sound levels are much too high. In some cases, it even exceeds the limit of what is considered safe for the hearing (Arbetsmiljöverket, 2010). Apart from damaging the hearing, high sound levels can also contribute to concentration difficulties, fatigue and irritation. Studies have also shown that the production of stress hormones increases, and the heart rhythm as well as blood pressure are negatively affected by noise (Babisch, Fromme, Beyer, & Ising, 2001; Englund, 2000). From that perspective, the subject of hertz and decibels is naturally important for the health and well-being of students and teachers in our preschools and schools.

Regardless of the number of decibels that sets the limit for silence, or how many decibels might occur in our preschools and schools, there are many more aspects of silence that are worth emphasising and shedding light on. As I understand it, therefore, silence must be considered from more angles than only hertz and decibels. One such angle is the linguistic aspect.

Silence from a Linguistic Viewpoint

From a linguistic point of view, silence can be described using the following synonyms: stillness, calmness, soundlessness, fade away, end, silentium (Walter, 1995). The English term 'silence' comes from the latter—the Latin noun silentium, which originates from the Latin word silere [to be quiet] (Collins English Dictionary, 1992; Simpson & Weiner, 1989).

In purely linguistic terms, it is interesting to establish that different languages place different meanings in a word such as 'silence.' In English, for example, the phrase to be silent means not only the absence of speech, but the lack of any kind of sound. By way of comparison, the Polish verb milczec means to refrain from talking. The Polish language thus has a narrower definition of silence than English does. In order to describe forms of silence other than just the absence of the spoken word, Polish speakers use the word cisza, which denotes a general stillness (Jaworski, 1993). In Swedish, the word tystnad carries the meaning of an absence of sound, a kind of soundlessness, and that a person does not say anything but stays quiet (Natur och Kulturs Svenska Ordbok, 2001). The Swedish tystnad can therefore be compared more closely with English than Polish, which has two different words for silence; one for complete silence or soundlessness and another that is used to describe a person refraining from talking.

But to describe silence solely through linguistic investigations is not enough to provide a sufficient and satisfactory picture. Silence as a linguistic phenomenon is one aspect to ponder and reflect upon, but there are so many more aspects of silence that are worth considering. One such aspect, which has a close natural connection to the linguistic one, is the communicative. Communication is an essential part of education, and one important dimension of communication is silence.

Silence and Communication

It is important to create time and space for silence, not only in everyday conversation, but also during more formal, structured discussions, such as at school. Formal interviews are another example of structured conversation, such as those that occur within the framework of a research study. In such situations, we should consider whether it is really necessary to ask all of the questions. Instead of extra questions, stillness and quiet could be a more tactful way of encouraging a person to tell more about their story and what they would like to share (van Manen, 1990). It is during the silent moments that a person can contemplate and reflect. At this point, we can again stop and think about what the situation is often like in our schools. Are students given a chance to contemplate by being left alone in the silent moments that can occur when two or more people are talking, or are these moments hastily filled with more words? According to Bollnow, it is when a conversation gradually subsides, as more and more pauses occur and it finally relapses into silence, that a conversation has been completed. An eventual return to silence is a sign of a good conversation: 'And when the conversation finally does sink into silence, it is no empty silence, but fulfilled silence' (Bollnow, 1982, p. 46). What is true of the spoken word likely also applies to written texts—the text is completed when its readers sink into silence through the stillness of reflection.

To be unable to speak or make one's voice heard is not the same as being silent; a person is silent only when they are able to talk—that is, that they have something to say (Merleau-Ponty, 1996). Let us stop to digest that for a moment. A person is silent only when he or she has something to say. What is it like in our schools today? Many students are considered by others, or by themselves, to be quiet. Perhaps they are not given, or do not take, the silent moment that is needed in order to share in a discussion. They remain quiet, despite the thoughts and opinions raised in their minds by an ongoing discussion, and they continue to be silent, even though they know the answer to a question the teacher just asked. But are they truly silent, then? Using the reasoning that a person is silent only when they have something to say, it follows that, if students who do not make their voices heard are to be considered silent, they must have something to say. Most quiet students likely have an opinion to add to the discussion, or an answer to the question, but, for some reason, they choose not to express it, becoming quiet in the eyes of others (or themselves). Perhaps these so-called quiet students are indirectly silenced out of a fear of answering incorrectly, or they are worried about expressing an opinion that their classmates will laugh at and ridicule.

When discussing the communicative aspect of silence, some go so far as to assert that meaningful silence can occur only when two or more people are involved—never when a person is alone (Jaworski, 1993). This can be compared with talking, which can very well be done by a person who is alone. Pittenger, Hockett and Danehy (1960) state that: 'it only takes one person to produce speech, but it requires the cooperation of all to produce silence' (p. 88). Meaningful silence can thus take place only when two or more individuals are involved. So how should we understand this? Can one person by themselves not experience meaningful silence? In contrast to Jaworski, I wish to claim that they can. Whether or not a silent moment is meaningful to an individual can for the most part be determined only by that person themselves. Meaningfulness is a subjective experience, which should also be true of silence. Students have even expressed as much in previous research studies. These students emphasised, for example, the value—the meaningfulness—of sitting quietly by themselves during a school day (Alerby, 2004). What these students were referring to was silence when it is perceived as comfortable, or even longed for. Silence can, however, also carry with it unpleasant and, in some cases, perhaps even dreaded aspects, and in such cases, silence is far from appreciated. So it is important to be attentive to and aware of the nature of the silence.

WHAT IS THE NATURE OF THE SILENCE?

As mentioned previously, silence in one form or another is always a part of our daily lives; it is impossible to completely avoid it. The question, though, is how the silence is perceived; is it good or bad? Or, to put it another way, is the silence perceived as constructive or destructive? Sometimes, silence can be perceived as heavenly, something longed for; at other times, it can be a forced and extremely unpleasant experience. These two angles of silence constantly work together and interact with one another, and the boundary between constructive and destructive silence is often a tenuous one. By its very nature, then, silence consists of both positive and negative values. One researcher who considers this matter is Jensen (1973). Jensen discusses five separate effects or functions of silence, where each function has a positive and negative value. He describes functions of silence and their positive and negative aspects as follows: Silence can have a linkage function (it may bond two or more people, or it may separate them); silence can have an affecting function (it can heal, but it can also insult by opening up old wounds); silence can have a revealing or revelation function (it can give a person insight or knowledge about something, someone or themselves, but silence can also be used to withhold information); silence can have a judgemental function (it can signal approval and appear helpful, but it can also signal disagreements or disapproval); and silence can have an activating function (it can signal thoughtfulness or cognitive activity, but it can also signal mental inactivity).

Allow me, please, to now transfer the various functions determined by Jensen into educational settings at school. For example, by allowing creative and permissive forms of silence in the classroom, students can, individually or as a group, be given time to reflect on an issue, an event or a situation, which can, in turn, lead to a sense of camaraderie and cohesion amongst the students. If silence in the classroom affects students in a positive way, it can help to solve problems and dilemmas as an aid to reflection and contemplation. But silence in a classroom can also feel uncomfortable and forced, which can cause a student to feel isolated and separated from the rest of the class—instead of being part of the group, he or she is shunned. Students who are forced into an uncomfortable and perhaps frightening silence may feel humiliated and ignored. This is often seen in the bullying scenarios that take place every day at schools around the world (more on that later).

Allowing time for silence in an educational setting, however, can make the actual learning process easier. The student in question is given a quiet moment to reflect on the subject or task, which can lead to him or her gaining insight and developing an understanding in that area.

Silence in the classroom between students—or between the teacher and the students—can also be a way to demonstrate a position of power, which could lead to information being withheld, therefore hindering the intended learning. On the contrary, a student may come to learn something completely different to what was intended with the education. In this context, though, that is a discussion best left to another occasion.

By allowing silence in educational settings, those quiet moments can facilitate and support decisive action and meaningful activities. The silence that can occur when students are engaged and interested in their assignments in the classroom may reflect deep thinking and reflection on the assignment. If, for various reasons, the students are instead indifferent and uninterested, perhaps their silence is signalling that their thoughts are engaged in a completely different direction than intended. Because, although Jensen (1973) describes how silence can reflect mental inactivity, I believe that it is extremely rare for a student at school to be mentally inactive, if it ever happens at all. A student might, on the other hand, be thinking about something completely different to what the teacher intended when planning and carrying out their teaching.

Silence can thus have different functions, and I would now like to highlight how it can sometimes be used as an instrument of power.

SILENCE AND POWER

As stated above, one of the inherent facets of silence is power, a facet that is not always easy to recognise. Whether or not this power is explicit or implicit, Foucault (1978, p 101) emphasises that silence is a 'shelter for power.' Individuals and groups have been silenced throughout history. Religious and political movements, for example, have on many occasions exercised their power to silence people, and continue to do so to this day. But it can also happen that people themselves choose silence as a way to exercise power.

One segment of humanity that has been silenced throughout history in many different situations is women. Through the centuries, women as a group have been silenced in both religion and politics. According to Christian tradition, for example, women are expected to stay silent, as exemplified by the Apostle Paul's letter to the Corinthians, where women are admonished to be quiet—to stay silent:

> The women are to keep silent in the churches; for they are not permitted to speak, but are to subject themselves, just as the Law also says. If they desire to learn anything, let them ask their own husbands at home; for it is improper for a woman to speak in church. (First letter to the Corinthians, chapter 14, verses 34–35, The Bible, 2006)

I would like to highlight here that in this example, silence is interpreted as an absence of the spoken word. To refrain from speaking is the same as being silent, an assumption that might be considered to be completely correct, but is in one respect based on the theory that it is silent when human voices are not heard, which is in turn based on a reasoning about hertz and decibels. But there are other aspects to take into consideration when discussing silence. The expression cum tacent clamant—speaking silence—can again be used as an example. It was used as early as 62 BC. In modern times, Bateson (1987), Dickinson (2020) and others have emphasised that a quiet person conveys a message, and Dickinson goes so far as to claim that the messages that say the most are often the silent ones.

In political contexts, it was not until 1919 that it was decided to allow men and women to have equal voting rights in Sweden. 1921 was the first year in which women were allowed to vote in the Swedish parliamentary election (Rönnbäck, 2004). Sweden thereby became the first Nordic country to introduce women's suffrage. Up until that year, the voices of Swedish women were to some extent silenced in political dialogues, and that is still happening today in some countries and regimes where they refuse to allow women to vote in politics. These women cannot make their voices heard at all.

Historically, some regimes and political organisations have deliberately and strategically silenced people. Even in today's globalised society, totalitarian regimes exist that silence a great number of people. Those who have been silenced by the likes of oppression and power may eventually feel that they are lacking a voice and cannot therefore be heard because no-one is listening. Freire (1972) wrote about the culture of silence, where, after a long period of forced silence, people started to believe that they did not have a voice and therefore had no control over their situation. They felt that they were beyond the possibility of being able to affect the situation because they were not listened to. Weil (2005) expresses similar thoughts:

> The afflicted are not listened to. They are like someone whose tongue has been cut out and who occasionally forgets the fact. When they move their lips no ears perceive and sound. And they themselves soon sink into impotence in the use of language because of the certainty of not being heard. (p. 91)

To lose the ability to voice one's thoughts and opinions can be devastating on several fronts. It is not the evil of the wicked, but the silence of the good that is dangerous, according to Arendt (1958).

Silence can be forced upon us for various reasons, such as because of oppression, ignorance and/or someone exercising their power, as is often true in the case of bullies at school or elsewhere. Arendt's reasoning is relevant to the modern school and the problem of bullying that often arises there, since one aspect of bullying is silence. Silence is found not only as a component of the act of bullying itself—such as shunning, the withholding of information or condescending looks—but can also come from other people, who, despite having seen the bullying, choose not to act. This kind of silence is devastating for the victim as well as the bully. Nor does a victim of bullying—perhaps a student—always tell someone about it. Perhaps the bullied student does not say anything about their situation because they do not dare to tell it to the adult world, such as the teacher or their parents (Gunnarsdottir et al., 2015). Instead, the student in question becomes and remains silent. But it may be the case that silence emanates from the fact that the student has been either directly or indirectly silenced. There are many possible explanations as to why bullying causes silence, regardless of whether it is the victim or the bully who is silent, or if it is those who may witness the bullying but fail to act. Whichever may be the case, it is of utmost importance to pay attention to how silence is used and how it is expressed. 'Aspects of power and marginalisation in relation to silence are therefore important to recognise in education' (Alerby, 2019, p. 537).

Arendt talks about the danger of good people remaining silent; that is, failing to react and make their voices heard. She argues that, if silence prevails in such situations, then evil becomes so commonplace that people no longer think or care about what is actually happening. Bullying that is confronted with silence—that is, when it is allowed to go unchecked for an extended period of time—can lead to desensitisation, which may eventually result in acceptance (Forsman, 2003). To borrow Arendt's words, bullying risks becoming a banality (Alerby, 2012).

There are other situations at school where silence is used as an instrument of power. For example, in some cases, silence is used by the teacher as a strategic means of expressing power. According to Jaworski (1993), the silence of a teacher highlights their dominance over the students, and he believes that, amongst other things, teachers often use silence in an attempt to get the students' attention, to interrupt them, to show their dissatisfaction, to regain order in the classroom or to produce calm in a tense situation—a kind of teaching strategy.

The teacher has the power to determine who should speak, and therefore also controls the silence. Who gets the question that a teacher asks? Who is allowed to make their voice heard in a discussion? Who is made to wait? Who gets to listen instead of speaking? The answer to these questions may depend, at least in part, on the prevailing culture. It is therefore time to consider silence in the light of various cultural aspects—school cultures as well as various national and cultural differences. Allow me to start with the latter.

SILENCE IN THE LIGHT OF CULTURAL ASPECTS

Apart from the fact that silence has various connotations and uses within different political or religious affiliations, the way silence is viewed can also depend on cultural variations. In spite of often considerable dissimilarities in how silence is treated and valued, the proverb 'speech is silver, but silence is golden' is found in most languages (Holm, 1976). In some cultures, such as the American one, the art of expressing oneself verbally is connected closely with success on both the private and professional levels, and a person in a leading position is expected to maintain their position through the spoken word. The opposite can also be seen; that is, that in some cultures, status is instead expressed by not talking—being silent. The talking is done by subordinates, so the one who has the power needs only to show themselves (Hedquist, 2006).

In Japan, for example, silence is valued just as highly as the spoken word. It is considered important for pondering what has been said, providing an opportunity for contemplation and reflection (Archer, 2001). It is worth mentioning in this context that, from a cultural and communicative perspective, there is an ambivalent view connected with the use of silence in Japan. Silence, and how it is valued, can be divided into two parts, on the basis that we ourselves consist of two parts—an inside and an outside. The inside is associated with truthfulness and sincerity, which are symbolically located in the heart and stomach. The outside, on the other hand, is associated with the face, the mouth and the spoken word. From the outside come misconceptions, illusions and falsehood, whilst it is silence that expresses the inner truth (Lebra, 1987).

Cultural differences can cause some misunderstandings when silence is interpreted in different ways (Jaworski, 1993). That silence can lead to misunderstandings is not limited to occasions when people from different national cultures meet; it happens every day when we meet and associate with people in the same social environment as ourselves. This is something that we should be especially observant of, particularly in schools and other educational institutions and pedagogical settings.

The way silence is viewed can differ from school to school, often depending on the prevailing culture at those schools. Is it accepted when a student is quiet, or is their silence perceived as provocative? Does the teacher ensure that there is a time and place for silence, or is the space immediately filled in? If, for example, a student fails to express themselves even when it is expected that they should have something to say on a matter, their silence can be interpreted in more than one way: it could be a sign of ignorance, insecurity, shame, lack of intelligence or nonchalance. But their silence could also be due to thoughtfulness and that they need time and space for reflection and consideration before the silence is broken. The question is whether this is permitted. And the answer to that question often depends on the prevailing school culture. A culture and classroom climate could be found within an entire school or in a certain class or group of students where one or more students become silent out of a fear of answering incorrectly or expressing an opinion that might be met with derisive laughter or silence. The opposite can also occur—that is, that silence is a natural and appreciated part of everyday life at school. It is not only in the educational sphere that silence is sometimes valued highly; people might choose to be in silence, and it is now time to take a closer look at this self-imposed silence.

CHOOSING SILENCE

Silence can be self-imposed, and it may be chosen for a variety of reasons. For example, people may choose silence as a means of protest, or in an attempt to protect somebody from something. One of the trademarks of silence is loyalty, which can, in turn, lead to a person refraining from revealing someone or something in an effort to protect that person, object or event (Sörlin, 2004). Another reason for choosing silence could be in a search for peace and tranquillity so as to think and reflect.

Silence is also chosen by the followers of some religious movements, such as the Trappists, whose members completely refrain from talking for certain periods of time (Brown, 1993). Other examples of religious followers within the Catholic Church who regularly and daily remain in silence are those who belong to the Benedictine and Jesuit Orders. These followers are silent when privately meditating, which they do at least twice a day. Buddhist philosophy and religion (see, e.g., Harvey, 2000; Southwold, 1983) is another tradition that values silence, using it, for example, as an element in their quest to answer various ontological (metaphysical) questions, such as the nature of reality.

Maitland (2009) mentions some clear examples of self-imposed silence, which have also had important historical ramifications, as follows:

Gautama Buddha's silent meditation under the Bo Tree some time between 566 and 368 BCE, Jesus of Nazareth's forty-day solitary fast in the Sinai

Desert c. 33 CE and Muhammed ibn Abdullah's annual Ramadan retreat to Mount Hira, near Mecca, culminating in his revelations of 610 CE, are rather obvious cases in point. (p. 46)

In our post-modern society, various forms of retreat have become popular, mainly in connection with religious or spiritual activities. The word 'retreat' can be defined as a kind of withdrawal, a period of inner stillness, where people meet for a few days of silence. But silence in these contexts means refraining from talking and instead listening inwardly—a form of contemplation. It was through contact with Anglicans in the 1950s that the Church of Sweden became interested in retreats (The National Encyclopedia, 2010c). Non-Christian religions, too, use retreats in their activities—to be in silence. To personally choose to be in silence is one of several aspects I have been highlighting and discussing. But silence can be even more multifaceted, so I would like to now elucidate and consider this in the form of literal, epistemological and ontological silences.

THE MULTIFACETED NATURE OF SILENCE

Silence is a complex and multifaceted phenomenon possessing many different dimensions, not solely in pedagogical settings but also in daily life. Foucault (1978) was one who emphasised the multifaceted nature of silence: 'There is not one but many silences' (p. 27).

There are thus many aspects of silence. From a phenomenological[1] angle, van Manen (1990) believes that there is nothing as silent as the

---

[1] At the turn of the twentieth century, German philosopher Edmund Husserl (1859–1938) introduced the idea of phenomenology for his entire philosophical approach, thereby founding the development of contemporary phenomenology. Phenomenology as a science is primarily a specific method and a specific mindset. The basic principle that came to characterise phenomenology was formulated by Husserl, and it went like this: 'Back to the things themselves.' Husserl argued that it is by means of experience that we can access the things themselves. What Husserl was referring to was how the world is subjectively experienced by our senses, and his aim was to find ways to describe the meaning of these subjective experiences. In other words, it is a description of how the world is experienced by the people who live in it. Phenomenology as a movement, however, came to have several different orientations, one of which is life-world phenomenology. Life-world phenomenology was developed by existential philosophers such as Martin Heidegger, Jean-Paul Sartre, Hans-Georg Gadamer and Maurice Merleau-Ponty, and is based on a life-world philosophy without transcendental reduction. Sources: Bengtsson (1993), Husserl (1995), Gadamer (1975), Heidegger (1993), Merleau-Ponty (1996).

completely obvious, that which we take for granted. From that reference point, he argues that this is the reason that behavioural science research is both possible and necessary. In order to develop this reasoning, we therefore need to question the things we take for granted and ask ourselves what is actually happening in what appears to be happening.

Van Manen (1990) illustrates the multifaceted nature of silence by describing three variations: the literal, the epistemological and the ontological.

Literal silence is characterised by the absence of speech. In some situations, it is better to stay quiet than try to fill the silence with words. This is also true of the written word. For both the spoken and the written word, it is sometimes preferable to let things go unsaid or unwritten. The silent moments in conversation or written texts are just as important and speak just as loudly as the spoken or written word. Instead of explicitly writing everything out in a text, therefore, it is in certain situations better to let things go unsaid, or perhaps unwritten—to leave the text silent. Instead, the text as a whole contributes to a certain understanding and gives a certain message, and the silent spaces are just as important as the words that are written.

Epistemological silence is what confronts us when we stand face to face with the unspeakable, the indescribable. These are, for example, the occasions when we know more than we are able to express, or, to borrow Polanyi's words, tacit knowledge (Polanyi, 1958, 1969). Polanyi emphasises that it is unthinkable that knowledge could ever be absolutely expressed or articulated. Behind the spoken and written words, there are—according to him—rich domains of the unspoken that constantly beckon to us. We could take a witness to a bank robbery as an example. Perhaps he is unable to describe what the robber looked like, but he can still be helpful to the police in drawing up a photofit of the perpetrator. So we have knowledge on one level, but we are unable to access it with our linguistic skills. Merleau-Ponty (1995) also emphasised that there is something that exists beyond what is said, something that cannot be communicated verbally, and he calls it a silent and implicit language.

Van Manen (1990) makes a few different distinctions in epistemological silence: (i) In some situations, when we experience the inexpressible, something impossible to talk about, something that lies beyond our own verbal proficiency, something that we are not personally able to put into words, another person may be able to make the inexpressible communicable. This happens in situations where we are unable to say what we really mean, but someone else manages to put it into words, and our reaction is 'I couldn't have said it better myself.' (ii) Some experiences turn out to be inexpressible within the bounds of one type of discourse but can be expressed in the context of another kind of discussion. An example of this is when the behavioural sciences have difficulty giving a satisfactory description of how love is experienced, yet it can be portrayed excellently through poetry, music, paintings or other art forms. But we must be clear that these two discourses—the behavioural sciences and the arts—differ in terms of their epistemological bases. The notion of what true knowledge is (and whether it even exists) and how we arrive at it can thus differ according to the type of discourse we are in (Gadamer, 1975). (iii) What may seem impossible to talk about in one situation may be expressed quite freely in another. Van Manen (1990) says that, in some situations, we can surprise ourselves by managing to express an experience that we previously were unable to put into words; neither verbally nor in writing: 'Did I really just say that?' 'Did I actually write that?'

Finally, Ontological silence is characterised by 'The silence of Being or Life itself' (van Manen, 1990, p. 114). According to van Manen, it is this form of silence that we turn to when we find ourselves in a difficult situation, in a similar way to how we return to silence after reading something or listening to a talk or a lecture. 'It is indeed those moments of greatest and most fulfilling insight or meaningful experience that we also experience the "dumb"—founding sense of silence that fulfils and yet craves fulfilment' (van Manen, 1990, p. 114). Another way of putting it is that the silence of being is a closeness to the truth. All speech emanates from silence, and the spoken word also falls back to silence (Bollnow, 1982). Genuine silence is accomplished by its transcending the spoken word.

*   *   *

This chapter has highlighted and discussed several aspects of silence. Our consideration has demonstrated the enormous range that silence covers—everything from hertz and decibels to silence from a linguistic and communicative standpoint. Other aspects include the relationship between silence and power, cultural perspectives and the desire for silence. Literal, epistemological and ontological silence are further examples of how silence can be understood and explained.

We humans perceive and understand silence in different ways. Silence may be desired in one situation, but in another, it can feel frightening or discomforting. People can even experience the same situation in completely different ways. People have different levels of sensitivity, so to speak, when it comes to silence. A silent person can, in certain situations, force another to talk. The silence perhaps makes the person in question feel uncomfortable and that they must break that silence by saying something. Another person, though, may be completely comfortable with the silence that can occur between people. Sometimes, we humans use sound to hide our need for silence. It is also the case that not all people are able to easily express themselves verbally. It is therefore time to take a closer look at some alternatives to the spoken word—alternatives in the form of art.

REFERENCES

Alerby, E. (2004). The appreciation of a quiet place at school. *Didaktisk tidskrift,* *14*(1), 57–61.

Alerby, E. (2012, May 2–4). *Rethinking the schoolyard as a place for silence.* Full reviewed paper presented at the conference Philosophical Perspectives in Outdoor Education, University of Edinburgh, Scotland.

Alerby, E. (2019). Places for silence and stillness in schools of today: A matter for educational policy. *Policy Futures in Education, 17*(4), 530–540.

Arbetsmiljöverket. (2010). www.av.se. 25 July 2010.

Archer, A. (2001). *A beginner's guide to Japan.* www.shinnova.com/part/99-japa/abj17-e.htm. 7 Dec 2001.

Arendt, H. (1958). *The human condition.* Chicago: Chicago University Press.

Arlinger, S. (1995). Det utsatta örat [The vulnerable ear]. In H. Karlsson (Ed.), *Svenska ljudlandskap. Om hörseln, bullret och tystnaden* [Swedish soundscape. About the hearing, the noise and the silence]. Göteborg, Sweden: Bo Ejeby Förlag.

Babisch, W., Fromme, H., Beyer, A., & Ising, H. (2001). Increased catecholamine levels in urine in subjects exposed to road traffic noise: The role of stress hormones in noise research. *Environment International, 26*(7–8), 475–481.

Bateson, G. (1987). *Steps to ecology of mind.* Northvale, NJ: Jason Aronson Inc.

Bengtsson, J. (1993). *Sammanflätningar. Husserls och Merleau-Pontys fenomenologi* [Intertwinings. The phenomenology of Husserl and Merleau-Ponty]. Göteborg, Sweden: Daidalos AB.

Bergmark, U., & Kostenius, C. (2012). Student visual narratives giving voice to positive learning experiences – A contribution to educational reform. *Academic Leadership Journal, 10*(1), 1–17.

Bollnow, O. F. (1982). On silence – Findings of philosophicopedagogical anthropology. *Universitas, 24*(1), 41–47.

Boverket. (2011). *Vad är ljud och buller?* [What is sound and noise?] www.bover-ket.se/Planera/planeringsfragor/Buller/Vad-ar-ljud-och-buller/. 9 Feb 2011.

Brown, L. (red.). (1993). *The new shorter Oxford English dictionary. On historical principles.* Oxford: Oxford University Press.

Buber, M. (1993). *Dialogens väsen: traktat om det dialogiska livet* [The essence of dialogue: Treaty on dialogue life]. Ludvika, Sweden: Dualis.

Collins English Dictionary. (1992). Glasgow, Scotland: Harper Collins Publishers.

Dickinson, E. (2020). Emily Dickinson. https://www.poetryfoundation.org/poets/emily-dickinson (accessed 2 July 2020).

Englund, A. (2000). *Trafikstress. En redovisning av utförande och resultat i KFB-finansierade forskningsprojekt 1996–1999* [Traffic Stress. A description of the performance and results in the KFB funded research 1996–1999]. Stockholm: Fritzes.

EUdict dictionary. (2019). Cum tacent clamant. http://eudict.com. 3 Nov 2019.

Foucault, M. (1978). *The history of sexuality: Volume 1: An Introduction.* New York: Pantheon Books.

Forsman, A. (2003). *Skolans texter mot mobbning – reella styrdokument eller hyll-värmare?* [School texts against bullying – Real policy documents or shelving heaters?]. Doctoral thesis, Luleå University of Technology, Luleå.

Freire, P. (1972). *Pedagogy of the oppressed.* London: Penguin Books.

Gadamer, H.-G. (1975). *Truth and method.* New York: Seabury.

Gunnarsdottir, H., Bjereld, Y., Hensing, G. & Petzold, M. (2015). Associations between parents' subjective time pressure and mental health problems among children in the Nordic countries: a population based study. BMC Public Health, 15(353). https://doi.org/10.1186/s12889-015-1634-4.

Harvey, P. (2000). *An introduction to Buddhist ethics: Foundations, values and issues.* Cambridge, MA: Cambridge University Press.

Hedquist, R. (2006). Lyssna och lära [Listen and learn]. *Aktum, 6,* 24.

Heidegger, M. (1971). *On the way to language.* San Francisco: Harper Collins Publisher.

Heidegger, M. (1993). *Varat och tiden* [Being and time]. Göteborg, Sweden: Daidalos.

Hellberg, A. (red.). (2002). *Buller och bullerbekämpning* [Noise and noise control]. Stockholm: Arbetsmiljöverket.

Holm, P. (1976). *Bevingade ord* [Winged words]. Stockholm: Albert Bonniers förlag.

Husserl, E. (1995). *Fenomenologins idé* [The idea of phenomenology]. Göteborg, Sweden: Daidalos.

Jaworski, A. (1993). *The power of silence. Social and pragmatic perspectives.* London: Sage.

Jensen, V. (1973). Communicative functions of silence. *ETC, 30,* 249–257.

Lebra, T. S. (1987). The cultural significance of silence in Japanese communication. *Multilingua, 6*(4), 343–357.

Maitland, S. (2009, orig. 2008). *A book of silence. A journey in search of the pleasures and powers of silence.* London: Granata Books.

Merleau-Ponty, M. (1995). *Signs.* Evanston, IL: Northwestern University Press.

Merleau-Ponty, M. (1996). *Phenomenology of perceptions.* London: Routledge.

Natur och Kulturs Svenska Ordbok. (2001). Stockholm: Publishing House Natur och Kultur.

Pittenger, R. E., Hockett, C. F., & Danehy, J. J. (1960). *The first five minutes: A sample of microscopic interview analysis.* Ithaca, NY: Paul Martineau.

Polanyi, M. (1958). *Personal knowledge.* Chicago: University of Chicago Press.

Polanyi, M. (1969). *Knowing and being.* Chicago: University of Chicago Press.

Rönnbäck, J. (2004). *Politikens genusgränser: den kvinnliga rösträttsrörelsen och kampen för kvinnors politiska medborgarskap 1902–1921* [Political gender boundaries: The women's suffrage movement and the struggle for women's political citizenship 1902–1921]. Doctoral thesis, Atlas, Stockholm.

Simpson, J.A., & Weiner, E.S.C (red.). (1989). *The Oxford English Dictionary, second edition, Volume XV, Ser–Soosy.* Oxford: Clarendon Press.

Sörlin, S. (2004). *Världens ordning* [The order of the world]. Stockholm: Publishing House Natur och Kultur.

Southwold, M. (1983). *Buddhism in life: The anthropological study of religion and Sinhalese practice of Buddhism.* Manchester, England: Manchester University Press.

The Bible. (2006). Bibelkommissionens översättning [Bible Commission translation]. Örebro, Sweden: Publishing House Libris.

The National Encyclopedia. (2010a). Tystnad, [Silence], www.ne.se.proxy.lib.ltu. se/sve/tystnad?i_h_word=tystnad. 26 June 2010.

The National Encyclopedia. (2010b). Ljud, [Sound]. www.ne.se.proxy.lib.ltu.se/ lang/ljud/243395. 24 July 2010.

The National Encyclopedia. (2010c). Retreat. www.ne.se.proxy.lib.ltu.se/lang/ retreat. 23 July 2010.

van Manen, M. (1990). In State University of New York Press (Ed.), *Researching lived experience. Human science for an action sensitive pedagogy.* London.

Walter, G. (1995). *Bonniers synonymordbok* [Bonniers thesaurus]. Stockholm: Publishing House Bonnier Alba AB.

Weil, S. (2005). *Attention and will. An anthology.* London: Penguin Books.

CHAPTER 2

Silence and the Arts

Thoughts and experiences can, of course, be expressed in a great variety of ways. Language takes on many different forms, including various symbolic systems—both visual and auditory. The question is whether (and in that case, how) these different languages are taken into account in the educational settings of our schools and elsewhere. I will therefore discuss some examples of nonverbal modes of expression, or, to put it another way, some of the arts will be highlighted. I will also reflect on the significance of silence and the role it plays in these art forms.

But let us first take a brief look at our Western culture. For quite some time, monomodality—that is, the use of a single type of sign in communication—has been clearly and expressly advocated and preferred. The most highly valued genres of written text, such as fiction novels, academic texts, official documents and various types of report, have been written almost exclusively without the use of images or illustrations (Kress & van Leeuwen, 2001). The visual arts have also been characterised by monomodality. Historically, artists predominantly used to employ the same methods and materials—oil paintings on canvas. Uniformity was also seen at concerts, where musicians often moved and were clothed in a uniform manner. Only the conductor and any soloist who might appear could deviate a little (but not too much) from the norm.

Even the theoretical and reviewing disciplines that developed for discussing these arts were uniform and monomodal. There was one language

© The Author(s) 2020
E. Alerby, *Silence within and beyond Pedagogical Settings*,
https://doi.org/10.1007/978-3-030-51060-2_2

for talking about visual art, another for talking about music, another for talking about written texts and so on. Each one of these languages had their own method, their own assumptions, their own subject-specific vocabulary and their own strengths and weaknesses (Kress & van Leeuwen, 2001).

In recent times, however, this monomodality has been somewhat loosened; what was previously dominated by monomodality can be said to have changed direction, trending more towards multimodality. This is noticeable in the mass media as well as in documents produced by businesses and organisations, universities, government offices and ministries. Within the various arts, too, such as the visual arts, music, theatre and dance, there is a clear trend towards crossing traditional boundaries and using a mixture of materials, methods and styles—an effort towards multimodality (Kress & van Leeuwen, 2001).

Monomodality and multimodality aside, the arts can often be regarded as nonverbal. In this chapter, I will highlight some of these nonverbal modes of expression: the written word, including poetry; texture; images; music; pantomime and mime. I intend to reflect on and problematise these modes of expression in relation to pedagogical settings and silence. In these symbolic systems, or nonverbal languages, silence can describe the form as well as the content; the modes of expression can in themselves be silent (the form) in the same way that what they convey (the content) is silent by nature. I will return to this when we elucidate and discuss these various modes of expression a little later.

THE WRITTEN WORD

In today's society—and more specifically, in today's schools—the written word is highly valued. Writing transforms language into a form of narrative discourse, and this discourse differentiates the narrative from silence: 'Narrative is discourse, and the prime rule of discourse is that there be a reason for it that distinguishes it from silence' (Bruner, 1996, p. 121). Writing can be described as a more deliberate process than the spoken word, which allows greater opportunity for reflection and contemplation (Dysthe, 1993). The written word can be seen on paper, and we can stand alongside it, so to speak, and behold it from the outside. The written word also endures in a completely different way to the spoken word; so we could say that written text leaves a lasting impression, whilst the spoken word is transient by nature. Additionally, a written composition appears to result in the writer gaining a higher level of awareness (Nygren & Blom, 1999), and writing can be considered the highest form of symbolic thinking (Vygotsky, 1978). By means of writing, it is possible for the author to separate themselves from the life-world, whilst simultaneously coming nearer to it: 'writing distances us from the lifeworld, yet it also draws us more closely to the lifeworld' (van Manen, 1990, p. 127).

Writing takes on great variety of forms. Poetry represents one of these that may be worth highlighting in connection with silence. Poetry came about in order to pass beyond words, and it is able to exceed the limits of language by evoking that which is impossible to articulate (Eisner, 1997). The very essence of poetry seems to be behind these words by van Manen: 'even in the most profound and eloquent poem it seems that the deep truth of the poem lies just beyond the words, on the other side of language' (1990, p. 112). But it is important to be aware that it is from silence that the words, and thus the text, emanate. Van Manen says that there is thus a similarity between poetry and phenomenology. Poetry and phenomenological texts speak partly through silence—they intend to convey more than what is explicitly stated. Both poetic and phenomenological texts intend to be silent at the same time as speaking—they are to a certain extent silent whilst also being explanatory. Reading poetry and phenomenological texts therefore requires sensitivity and openness, a responsiveness to the silence surrounding the words.

Poetry can also be silent as to its content. The Polish poet Anna Kamienska valued silence in her own poems, and she wrote about silence as an opportunity, or perhaps even a necessity, for communication. She felt that silence in human communication is universal and more effective and potent than the spoken word. To illustrate her point and show her confidence in silence, she symbolically included a few blank pages in one of her poetry collections (even though her publisher never agreed to it) (Jaworski, 1993); in this way, the poetry became silent in form.

Another poet who experiments with silence in both form and content is the Swiss German-speaking Eugen Gomringer. One of his stated goals is precisely to unite form and content, and he was one of the key figures of the German-language concrete poetry movement, which developed worldwide in the 1950s and 1960s (Solt, 1968). In 1954, Gomringer wrote his famous poem *Silencio* [Silence] (see below).

silence	silence	silence
silence	silence	silence
silence		silence
silence	silence	silence
silence	silence	silence

Concrete poetry is, according to Bann (1967), symbolic in regard to the theme of the poem in two ways. In Gomringer's *Silencio*, or *Silence*, the most obvious and significant way this is shown is through the empty—or silent—space in the middle of line three. The empty space represents an extended, silent pause that occurs when the poem is read aloud. The second is the repetitive nature of the poem; according to Jaworski (1993), undifferentiated repetition borders on silence. Such repetition can eventually lead to a text or speech becoming mute—the repetitions prevent the message from being heard. It transforms, instead, to silence, just as the constant dripping of a tap is eventually not heard any more—the drops fall into silence. To connect this with the school world—and, for that matter, other educational settings—we can think about a teacher who constantly repeats the words 'You can be quiet now' to a group of boisterous and talkative students who neither 'hear' nor follow the teacher's request. The teacher's words end up being repeated into silence. Returning to Gomringer's poem, we can add that the only word he uses in the composition is 'silence.'

Silence is therefore explored by poetry. But silence is also both evident and necessary when words are used in everyday life. Whether spoken out loud or written down, words without silent spaces between them would make communication barely possible. As an experiment, allow me to give a practical example of this by rewriting the previous few sentences whilst omitting the silent spaces between the words.

> But silence is also both evident and necessary when words are used in everyday life whether spoken out loud or written down words without silent spaces between them would make communication barely possibles an experiment allow me to give apractical example of this by rewriting the previous few sentences whil stomitting the silent spaces between the words.

The example above, with words arranged one after the other without any silent spaces, speaks its own silent language—that silence is an indispensable component of the language we use every day. For most of us, this is so obvious that we do not stop to think about it until the notion is challenged. The fact that this is obvious should not be interpreted as meaning that the silence, or empty spaces, between words lacks meaning. Instead, the opposite is true: it is the silence that gives meaning to the message. It could be said that each word is connected to the next one through the silent space between them in a continuous flow. In connection to this argumentation I would also like to emphasise that a word needs to be in solitude, but often together with other words—alone together—to give meaning to the text (Alerby, 2019b).

A further significant aspect of silence is that, when words are lacking, silence can in some circumstances serve as a refuge or protection. This is expressed by Jaworski (1993) as follows: 'When words fail poets, when artists find language inadequate to express themselves, they find refuge in silence' (p. 161). Another diametrically opposed strategy is to produce even more words when there is actually a lack of meaning behind them. The one who does not have anything to say—whether in speech or writing—can easily become verbose.

Regardless of what form of written expression is used—from poetry to everyday written language—something they all have in common is that the written word does not emit any sound and could therefore, from one standpoint, be considered silent. On the other hand, the written word does convey a message of some form, so it could be said that the text is speaking to the reader. Whether or not written language is considered to be silent or as some form of language that speaks to us depends on how silence is viewed. If silence depends on not exceeding a certain number of decibels, then the written word can be considered silent. If, however, silence is viewed as something other than the lack of auditory stimulation and it is instead what is conveyed that is important, we might say that written text speaks to its readers.

In the school environment, the written word is often highly valued, and a great deal of time and effort is spent on the students' writing skills. Could it be the written word is perhaps valued as the highest form of communication in the school world due to its enduring nature? Thanks to their permanence, the words written in, for example, an essay or written test can be read several times, assessed and graded. The words, which remain in the form of graphite particles on a piece of paper or as laser-printed letters, are there as a kind of evidence in case there should ever be a discussion about how a written text has been assessed or graded.

Although the written word is not completely everlasting, as a contrast to its relative permanence, consider the spoken word. Words that are expressed as answers to a teacher's questions or as an oral review for a school assignment quieten down and disappear into nothing, even if their message may long be etched into the minds of their listeners. Returning to the written word, we can stop and reflect on how that which is not written, that which lies between the words—the silence—is taken into consideration in educational settings.

The argument has already been made that silent spaces are needed between both written and spoken words, but other forms of expression and art also need silent spaces. One of these is texture.

SILENCE AND SOLITUDE IN THE LIGHT OF TEXTURE

Texture is one of the elements of art and design, similar to, for example, shape and colour. It can be simply described as a surface quality, both tactile and visual, covering both nature and culture, and indeed much of life itself. As a professional term, texture is used, not only in arts and crafts, but also in disciplines such as music, language and gastronomy, although carrying slightly different meanings (Loan, 2002).

'Texture' has the same origin as the term 'text,' and means 'weave,' 'tissue,' 'spin' or 'joining.' Within all visual arts, as well as craft design and architecture, texture serves as a formal aesthetic tool in the same way as form and colour (Doseth Opstad & Alerby, 2017). Put simply, texture is the visual and tactile character of surfaces and is discerned through the senses and perception, and we use a variety of expressions for describing textures, like 'soft,' 'hard,' 'smooth,' 'rough,' 'spiky,' 'uneven' and 'furry' (Opstad, 1990, 2010).

Absolutely everything that surrounds us has surfaces containing different textures; all the artefacts, things in nature, the animals and humans, although we cannot always feel it or see it. The earth's surface is a patchwork of different textures. The same applies to mankind itself—the human body, its skin and hair all consist of a variety of textures. Everywhere in our daily lives, we find different textures in the clothes we wear, the things we surround ourselves with and especially in nature. Since we are all surrounded by, and surround ourselves with, textures from when we come into this world until the moment we die, we might think that texture is given more attention than it does.

But what does texture have to do with silence? The example given above, where words were written together in a single sentence without spaces between the words, appears as a completely different texture, and a completely new understanding of the text arises along with it. It therefore becomes clear that text, like a texture, exists as a combination of many smaller elements—elements of texture that come together to form a whole. Waterhouse (2013) argues that there becomes only one uniform surface, a single texture, when the viewer perceives the elements as merging into a whole. The importance of silent spaces applies to all types of texture (and texts). All of them need spaces because, without spaces, there is no texture (or readable text). And these spaces can, in turn, be considered silent. Figure 2.1 shows an example of the importance of silent spaces—spaces-in-between—in order for a texture to actually be a texture. It is also an example of the repetitive nature of texture. The spaces-in-between in textures can also be viewed as cracks that unleash dissimilarities and disparities, in doing so, enables something new, or something

Fig. 2.1 An example of the significance of silent spaces between elements of a texture. It is also an example of the repetitive nature of a texture. (Photo by Kari Doseth Opstad)

different and contrasting. As Leonard Cohen (1992) sang in 'Anthem': 'There is a crack, a crack in everything. That's how the light gets in.'

Additionally, in the same way that a word in a text needs to be in solitude, but often together with other words, a part of a texture needs to be in solitude, but together with other texture parts (Alerby, 2019b).

In (art)educational settings, texture is essential for both visual and tactile attention, as well as aesthetic compliance, aspects that are significant for exploration and creative activities (Alerby, 2019a, 2019b; Doseth Opstad & Alerby, 2017).

Visual Art as a Form of Expression

As previously argued, language consists of much more than merely the spoken and written word. It can, for example, also include pictures and paintings (Dewey, 1991). There is something that lies beyond what is said—something that is unable to be conveyed verbally—and it is this unspoken language that can be illuminated through the creation of an image (Merleau-Ponty, 1995). A person who expresses themselves by creating something, such as by painting or drawing, thinks through the body's senses (Arnheim, 1969). By viewing an image, it is possible to find meaning and messages that the picture expresses; the image is saying something to its observers. A picture or drawing can therefore be seen as an expression of a symbolic language, which can be a methodological aid in making people's experiences, opinions and observations visible (Alerby, 2015).

The content of a picture, a painting or a canvass can also be silent. Examples include the monochrome paintings created by Yves Klein and Robert Rauschenberg, amongst others. French artist Yves Klein was born in 1928 in Nice, and died at just 34 years of age in Paris. Klein was both a painter and sculpturer, and he can also be described as a kind of performance artist. He is considered to be one of the greatest figures of post-war avant-gardism. He is most known for his large, monochrome paintings in International Klein Blue (IKB), a colour he created himself.

Robert Rauschenberg was an American artist who lived between 1925 and 2008. He is perhaps best known for his *Monogram* artwork, or, as it is commonly referred to, 'The Goat'—a stuffed goat with a car tyre around its belly, which can be viewed at the Modern Art Museum in Sweden, alternating between Stockholm and Malmö. It is, however, his white monochrome paintings that are of interest in the context of silence.

It was during the early 1950s that Rauschenberg painted a series of images that were only white. He later went on to also paint a series in black. Both Klein's and Rauschenberg's monochrome paintings are in some way reminiscent of sound and silence, or movement and stillness, but they can also be regarded as containing an emptiness (Jaworski, 1993).

Music: Can It Be Considered Silent?

Still another mode of expression worth highlighting in this context is music. It might at first seem somewhat strange to make this connection, since music, by nature, is not usually considered soundless, or silent. Nevertheless, silence is actually often regarded as an important element of music—it gives the mind room to experience the moment or the meaning behind the notes. These experiences are to be considered both personal and unique to each individual. Composers, musicians and conductors often spend considerable time thinking about how long the silence should be between the notes or percussion sounds in a piece of music in order to give the right balance to the work. The basic prerequisites for all music are sound and silence, which Achino-Loeb (2006) expresses as follows: 'Silence is the necessary ingredient for our experience of music because it marks the boundaries of what we consider music, for musical sound is such only when framed by rests and pauses' (p. 5).

Music can be regarded as the language of the soul, a voice for the inexpressible (Barnard, 1913). Bourdieu expresses similar thoughts when he says that music places a person alongside the words. His thinking is expressed as follows: 'Music is hand-in-glove with the soul: there are innumerable variations on the soul of music and the music of the soul' (Bourdieu, 1993, p 103).

There is also the well-known piece called 4′33″ by American composer John Cage. In this piece of music written for the piano, no music is played—it is silent (see Fig. 2.2).

Instead of listening to a piece performed by a pianist, the audience's attention is shifted to any other sounds that may reach them during the

Fig. 2.2 John Cage's score for 4'33". The original has disappeared, but it was recreated in 1989 by David Tudor (Gann, 2010, pp. 178–180)

4 minutes and 33 seconds that the composition takes to perform. Cage uses this silence as the main component of his work and concludes that silence is greater than sound (Cage, 1961, 1997). At the same time, he lets the audience understand that absolute silence does not exist, that there is

always some kind of sound that can be heard—sound that is worth high-lighting in connection with music. Cage wrote 4′33″ when he discovered that there is no absolute or total silence, and his inspiration for the work came from seeing Rauschenberg's white canvas (Achino-Loeb, 2006; Jaworski, 1993).

4′33″ was first performed on 29 August 1952 by the young pianist David Tudor at Woodstock in New York, causing a scandal. Cage himself said afterwards:

> People began whispering to one another, and some people began to walk out. They didn't laugh—they were just irritated when they realized nothing was going to happen, and they haven't forgotten it 30 years later: they're still angry. (Kostelanetz, 1988, p. 66)

Although 4′33″ is often described as Cage's silent piece, Gutmann (1999) believes that it is not silent at all. Whilst the pianist strives to be as silent as possible during the performance, Cage breaks with tradition by shifting the focus away from the pianist and the instrument on the stage towards the audience and even beyond the concert hall itself. As the audience's attention changes direction, they become aware of many different sounds—a chair creaks as a person changes position, someone else coughs, a lorry drives past on the road outside, the air conditioning hums, a door squeaks as it opens and so on.

Cage argued that silence as a concept must be redefined, and he claimed that there is no objective dichotomy between sound and silence. The essential meaning of silence is the giving up of intention, argued Cage, and he emphasised that silence is simply the absence of intended sounds, or, to put it another way, to turn off our awareness (Kostelanetz, 1988). These ideas mark the most important turning point in Cage's compositional philosophy.

Another piece of music composed by Cage that is connected with silence is Organ2/ASLSP (As SLow aS Possible). The basic idea is to determine how slowly the piece can be played. ASLSP was originally written in 1987, and the piece usually took between 20 and 70 minutes to perform. At a conference in 1997, musicians and philosophers began to discuss the implications of Cage's instructions to play ASLSP as slowly as possible. This eventually resulted in a project where an organ performance, which is currently under way, is planned to last for 639 years. The music is being performed in the St. Burchardi church in Halberstadt, Germany, and started with a pause on 5 September 2001 that lasted for seventeen months, during which time there was silence. The first sound (that belonged to the music) was heard on 5 February 2003, and the entire performance is expected to be completed in 2640 (Halberstadt event, 2011).

Pantomime and Mime: Theatre Without Words

Silence also has a direct importance for other art forms. Pantomime and mime are forms of theatre or acting that are mute, or silent, if you will. The terms *pantomime* and *mime* come from the ancient Greek and Roman theatres. Pantomime often portrays a more serious kind of drama, whilst mime is most commonly associated with what we today call slapstick, satire and parody (Boucher, 2002). Pantomime and mime are performed using the body and facial expressions, but no words are spoken; instead, it is body language that is used. It is a kind of mute acting that speaks mainly to the emotions rather than the ears. Silence is a natural and expected part of these arts. In contrast, actors in a theatre play mostly use their voices. How, then, is silence considered and used in these performances?

Regardless of whether the performance is based on pantomime, mime or regular theatre, it can be pointed out that it is not solely the spoken word that bears the message. Body language in the form of facial expressions and posture has a significant and important role in conveying the message. Silence can be used to emphasise a message, not only in pantomime and mime but also in regular theatre; for example, through the use of silent intervals. Silence can thus be viewed as an important complementary dimension to the message.

*   *   *

In summary, we can say that nonverbal modes of expression, such as poetry, images, music or pantomime, convey a message. Meanwhile, the message can be interpreted in a variety of ways (as is also true of the spoken or written message), and it is not always so easy to fully express it orally—that is, it is not always simple to talk about it. One reason for this is that many art forms transcend words (Bourdieu, 1993).

In pedagogical settings, however, it is important to be observant and responsive to students (or teachers, for that matter) varying as to how proficient they are in articulating themselves using different modes of expression. Since language contains much more than the spoken or written word, it is important that other means of communication are allowed to be expressed—all so that what is hidden behind the spoken word can have an opportunity to be heard. In order that, not only what lies beyond the spoken word, but also that which is explicitly said is heard and really listened to, I will use the next chapter to discuss the relationship between listening and silence.

*   *   *

REFERENCES

Achino-Loeb, M.-L. (red.). (2006). *Silence. The currency of power.* New York: Berghahn Books.

Alerby, E. (2015). 'A picture tells more than thousand words.' Drawings used as research method. In J. Brown & N. Johnson (Eds.), *Children's images of identity. Drawing the self and the other.* Rotterdam, Netherlands: Sense Publisher.

Alerby, E. (2019a, March 5–6). *Texture: In the light of senses, silence and the lived body.* Invited paper at the pre-conference "Educating the Senses in a Globalized World", The Nordic Society for Philosophy of Education, Uppsala University, Sweden.

Alerby, E. (2019b, April 10–12). *Silence, senses and solitude in the light of texture.* Paper presented at the International Pandisciplinary Symposium on Solitude in Community – Alone Together, York St John University, York, UK.

Arnheim, R. (1969). *Visual thinking.* Berkeley, CA: University of California Press.

Bann, S. (1967). *Concrete poetry: An international anthology.* London: London Magazine.

Barnard, W. F. (1913). *The tongues of toil and other poems.* Chicago: The Workers' Art Press.

Boucher, T. (2002). *Pantomimteatern.* www.pantomimteatern.com/om-oss/historia/mimkonsten/. 19 Jan 2011.

Bourdieu, P. (1993). *Sociology in question.* London: Sage.

Bruner, J. (1996). *The culture of education.* Cambridge, MA: Harvard University Press.

Cage, J. (1961). *Silence: Lectures and writings*. Middletown, CT: Wesleyan University Press.

Cage, J. (1997). *I-VI John Cage*. Hanover, NH: University Press of New England.

Cohen, L. (1992). 'Anthem' in the music album *The Future*. https://www.leonardcohen.com/. Accessed 16 Jan 2020.

Dewey, J. (1991). *How we think*. New York: Prometheus Books.

Doseth Opstad, K., & Alerby, E. (2017). *Textur, tystnad och kroppslighet* [Texture, silence and embodiment]. Paper presented at the e17 conference: 'Tacit knowing or just plain silence?', 31 October–2 November 2017, Umeå, Sweden.

Dysthe, O. (1993). *Writing and talking to learn. A theory-based, interpretive study in three classrooms in the USA and Norway*. Tromsø, Norway: University of Tromsø.

Eisner, E. W. (1997). The promise and perils of alternative forms of data representation. *Educational Researcher, 26*(6), 4–10.

Gann, K. (2010). *No such thing as silence. John Cage's 4'33"*. London: Yale University Press.

Gutmann, P. (1999). *John Cage and the Avant-Garde: The sounds of silence*.www.classicalnotes.net/columns/silence.html. 20 Jan 2011.

Halberstadt event. (2011). www.john-cage.halberstadt.de. 22 Jan 2011.

Jaworski, A. (1993). *The power of silence. Social and pragmatic perspectives*. London: Sage.

Kostelanetz, R. (1988). *Conversing with Cage*. New York: Limelight.

Kress, G., & van Leeuwen, T. (2001). *Multimodal discourse. The modes and media of contemporary communication*. London: Arnold Publishers.

Loan, O. (2002). *The elements of design: Rediscovering colors, textures, forms and shapes*. London: Thames & Hudson.

Merleau-Ponty, M. (1995). *Signs*. Evanston, IL: Northwestern University Press.

Nygren, L., & Blom, B. (1999). Analys av korta narratives [Analysis of short narratives]. In J. Lidén, G. Westlander, & G. Karlsson (red.), *Kvalitativa metoder i arbetslivsforskning* [Qualitative methods in working life research]. Uppsala, Sweden: Rådet för arbetslivsforskning.

Opstad, K. D. (1990). *Teksturer i vev* [Textures in tissues]. Master thesis, Statens lærerhøgskole i forming Oslo.

Opstad, K.D (2010). Estetisk dannelse – estetiske fags bidrag i skolens dannelsesperspektiv [Aesthetic education – Aesthetic subjects' contribution to the formation perspective of school]. In M. Brekke (Ed.), *Dannelse i skole og utdanning* [Formation in school and education]. Oslo, Norway: Universitetsforlaget.

Solt, M. E. (1968). *Concrete poetry: A world view*. Bloomington, IN: Indiana University Press.

van Manen, M. (1990). In State University of New York Press (Ed.), *Researching lived experience. Human science for an action sensitive pedagogy*. London.

Vygotsky, L. S. (1978). *Mind in society. The development of higher psychological process*. Cambridge, MA: Harvard University Press.

Waterhouse, A. L. (2013). *I materialenes verden; perspektiver og praksiser i barnehagens kunstneriske virksomhet* [In the world of materials; perspectives and practices in kindergarten artistic activities]. Bergen, Norway: Fagbokforlaget.

CHAPTER 3

What Is Heard in Silence? Or the Art of Listening

What do we hear in silence? To be able to answer that question, we need to listen—listen to the silence (if there is any). But what does it mean to listen? I shall endeavour to answer this question by highlighting and discussing the phenomenon of listening and its inherent relationship with silence.

The Swedish word *lyssna* [to listen] is defined by the *Natur och Kulturs Svenska Ordbok* dictionary (2001) as 'using the ears in order to hear,' and *The National Encyclopedia* (2010) explains the meaning as the 'deliberate use of the sense of hearing so as to perceive (and usu. differentiate) sound.' It is interesting to note that both the *Natur och Kulturs Svenska Ordbok* dictionary and *The National Encyclopedia* describe listening as hearing or using the sense of hearing, thereby equating listening with hearing.

Are they really the same thing? Although we may hear a person speak, we have to consider whether we always listen to what they are saying. I would argue that this is not the case. We humans can hear without listening. To truly listen to someone, we need to actively direct our attention and awareness to that individual. We therefore need to give the other person time and space to talk—time and space where silence should prevail. Silence is therefore needed when listening. We can thus give another person silence, but we can also take silence away from someone, which is an argument that finds parallels in Buber's (1978, 1988) thinking.

This swaying between giving and taking is key to Buber's view of education, amongst other things. On the one hand, Buber believes that students are not merely passive, requiring that teachers just fill the students with knowledge. But on the other hand, neither are students completely active, so that the teacher needs only to release the students' inner creative strength. Instead, the motivation for learning comes about when a teacher and student meet each other. This leads to students trusting and having confidence in their teachers, and teachers having trust and confidence that students will take advantage of their opportunity for learning and making progress—all in a constant positive feedback loop. In order for this to truly become reality, silent moments are needed, where the individuals in question are given opportunity to listen to what is said.

Adelmann (2008), too, argues that hearing is not the same as listening. Instead, he asserts that listening is a form of interaction, which can be expressed as a linguistic response. Listening makes a difference in both private and professional life, and even from a societal perspective. In the long term, this difference can change a person's life, says Adelmann. Questions that arise in this context are whether listening finds any place in school, and what role silence plays in relation to listening. Perhaps it is this that Schultz (2010, p. 1) emphasises: 'Understanding the role of silence for the individual and the class as a whole is a complex process that may require new ways of conceptualising listening.'

The Role of Listening in School

Basing his conclusion on philosophers Gemma Corrardi Fiumara and Roland Barthes, Adelmann (2008) states that our Western knowledge tradition tends to disregard the importance of listening, and listening simply does not exist as a recognised discipline. A review of various educational documents from when public school began in Sweden in 1842 until the year 2000 shows that listening has changed in meaning in the steering documents for schools. In the beginning, a vertical communication pattern was the norm. Students had to listen to the spoken and written words coming from society's authorities, and examples provided of these authorities are Luther's Small Catechism, a priest's sermon from Sunday service, the things taught by elementary school teachers and so on. This pattern has been transformed to a more horizontal and democratic communication pattern. Students in the twentieth century thus had greater opportunity to speak, and the spoken language was given greater emphasis in the steering documents; though it must be emphasised that, although the spoken language was given greater attention than previously, it was only on the part of the speaker. 'They learned to talk but were told to listen' (Adelmann, 2008, p. 48). It was, however, thanks to this that students were given greater opportunity to listen.

In previous steering documents for Swedish schools (Curriculum for the compulsory school system, preschool class and the leisure-time centre—Lpo94), listening was considered a communication skill that the school was expected to teach to students (Ministry of Education and Research, 1994). In the latest steering documents for schools (Curriculum for the compulsory school, preschool class and school-age educare—Lgr11), listening is also a central objective in some subjects, such as Mother Tongue, which includes the heading 'Speak, Listen and Converse' in all classes of compulsory school (Swedish National Agency for Education, 2011a). To ensure that students have acquired the art of listening, they are tested in national exams (Swedish National Agency for Education, 2011b). An example of this is the English subject test in year 9, which is entitled 'Listen and Learn.' These examples apply to Sweden, but parallels can be found in the schools and educational organisations of other Western countries.

Within the scope of this publication, I do not intend to delve deeply to ascertain how listening is treated in schools, but, according to Hedquist (2006): 'it can, however, be seen that teachers do work to train students to listen, though mostly in preschool' (p. 24).

Adelmann (2008), who highlights two approaches to listening in schools—the authoritarian versus non-authoritarian perception of knowledge—believes that teachers are thereby faced with didactic choices. Does the teacher want there to be one-way or two-way communication in the classroom? Does the teacher want to use a teacher-based or student-based approach to the education? Should questions and answers or conversations be used? Listening takes on different roles depending on how these questions are answered, but what we must especially ask ourselves is: what does this have to do with silence?

The Role of Silence in Listening

Silence is a natural part of listening, as Cooper (2012) emphasises: 'To be a good listener requires an ability to remain silent' (p. 54). If we are to be able to truly listen to what another person is saying, we ourselves need to be quiet. As the conversation progresses, silence and speech are successively given and taken. In conversing with others, we need to allow time for silence and quiet moments so that the other person can join the discussion. If this time is not given, or if the person cannot, does not want to or is not able to take that moment of silence and fill it with their own words, no conversation takes place.

Silence also allows the listener to think about and respond to what has just been said, so we should reflect on how we communicate with each other. This, in turn, means that silence is an important and often unavoidable part of human communication. But perhaps the importance of silence in our way of communicating with each other is neglected: 'silence is no doubt the most marginal aspect of linguistic action imaginable' (Verschueren, 1985, p. 74).

If a teacher has a positive attitude towards listening, and therefore really listens actively to those who are speaking, it becomes easier for others to express their thoughts, observations and experiences. When teachers themselves take a step back to allow room for silent pauses, students are given an opportunity to act and grow. This reasoning about communication and relationships can be linked with Lögstrup (1975) and what he refers to as the basic phenomenon of ethics, namely: 'to dare to go forward in order to be met' (p. 27).

THE RIGHT TO BE LISTENED TO

Article 12 of the United Nations Convention on the Rights of the Child (1989), which deals with freedom of opinion and the right to be heard, the obligation and importance of establishing a child's perspective is emphasised, and that 'the child shall in particular be provided the opportunity to be heard.' In order for this statement to be effective, we need to listen to the children. To truly listen to children and not only hear what they say is the most essential part of article 12 (Kellett, 2010).

A problem that can sometimes arise is that even when young people, such as students in a school class, say something and express an opinion, their voices are not always listened to (Smyth, 2006a, 2006b). Of course, this does not apply only to young people—there are other groups or individuals who are not listened to, whose voices are silent or silenced. It is not surprising that many students express that they do not feel part of the class because their voices are ignored or underappreciated—they are not listened to. Not being allowed to make their voice heard can, in turn, lead to a student becoming quieter and quieter. I would argue that the important thing in this context is that the teacher should be alert to silence and take it into consideration. A teacher's task is therefore to listen to both what is said and what is not said. However, school is not always a place where students feel they are listened to, which challenges teachers as well as other adults in the school to make room for the student's voices (Bergmark & Kostenius, 2009).

In pedagogical research, studies involving children are often conducted. Researchers may wish to study how children experience a certain phenomenon, such as something within the school. This type of research is an important element in ensuring that children's voices get heard—that is, listened to—and silence is important in achieving that.

* * *

In this chapter, I have highlighted and reasoned on the relationship between silence and listening. For people in general, and especially for teachers, it is important to really listen to the students. Silence is an important aspect of listening; and listening, in turn, is an aspect that is appreciated at school—two aspects that I have discussed above.

Silence is both an explicit and implicit part of the school world, and it is now time to take a closer look at how silence is often viewed in various pedagogical settings. What is the significance of silence in pedagogical situations? How is silence valued in the school world? These questions will be considered in the next chapter, 'The Significance of Silence in Pedagogical Settings.'

REFERENCES

Adelmann, K. (2008). *Konsten att lyssna. Didaktiskt lyssnande i skola och utbildning. The art of listening* [The art of listening. Didactic listening in school and education] Lund, Sweden: Studentlitteratur.

Bergmark, U., & Kostenius, C. (2009). 'Listen to me when I have something to say' – Students' participation in research for sustainable school improvement. *Improving Schools, 12*(3), 249–260.

Buber, M. (1978). *Between man and man.* New York: Macmillan Publishing Co.

Buber, M. (1988). *The knowledge of the man: Selected essays.* Atlantic Highlands, NJ: Humanities Press International, Inc.

Cooper, D. E. (2012). Silence, nature and education. In H. Fiskå Hägg & A. Kristiansen (Eds.), *Attending to silence. Educators and philosophers on the art of listening.* Kristiansand, Norway: Portal Academic.

Hedquist, R. (2006). Lyssna och lära [Listen and learn]. *Aktum, 6*, 24.

Kellett, M. (2010). Small shoes, big steps! Empowering children as active researchers. *American Journal of Community Research, 46*, 195–203.

Lögstrup, K. E. (1975). *Den etiske fordring* [The ethical requirement]. Köpenhamn, Denmark: Gyldendal.

Ministry of Education and Research. (1994). *Curriculum for the compulsory school system, pre-school class and the leisure-time centre—Lpo94.* Stockholm: Fritzes.

Natur och Kulturs Svenska Ordbok. (2001). Stockholm: Publishing House Natur och Kultur.

Schultz, K. (2010). After the Blackbird Whistles: Listening to Silence in Classrooms. Teachers College Record. The Voice of Scholarship in Education.

Smyth, J. (2006a). 'When students have power': Student engagement, student voice, and the possibilities for school reform around 'dropping out' of school. *International Journal of Education in Leadership, 9*(4), 279–284.

Smyth, J. (2006b). Educational leadership that foster students' voice. *International Journal of Education in Leadership, 9*(4), 285–298.

Swedish National Agency for Education. (2011a). *Del ur Lgr 11: kursplan i modersmål för grundskolan* [Part of Lgr 11: Syllabus for Mother Tongue for compulsory school]. www.skolverket.se/content/1/c6/02/21/84/Modersmal. pdf. 2 Feb 2011.

Swedish National Agency for Education. (2011b). *Om nationella prov* [About national tests]. www.skolverket.se/sb/d/2852. 2 Feb 2011.

The National Encyclopedia. (2010). Lyssna, [Listen]. www.ne.se.proxy.lib.ltu.se/sve/lyssna?i_h_word=lyssna. 22 July 2010.

United Nations Convention on the Rights of the Child. (1989). UN General Assembly Document A/RES/44/25.

Verschueren, J. (1985). *What do people say they do with words: Prolegomena to an empirical-conceptual approach to linguistic action.* Norwood, NJ: Ablex.

CHAPTER 4

The Significance of Silence in Pedagogical Settings

What is the significance of silence in various pedagogical settings, and how is the spoken word valued in the school world? By talking—using words—we can converse and discuss different things, concepts, phenomena and so on. The spoken word is therefore greatly important in order for us to be able to gain knowledge, insight and understanding about our world. It is thus crucial for both life itself and in various educational contexts.

> Thanks to the existence of words, we are able to discuss the world and our life. Words have become the tools of knowledge. Words make it possible for us to deliberate with each other on what exists and on what it is to be a human in the world. (Aasland, 2012, p. 249)

As previously shown, the spoken word (and the written, for that matter) requires silence—silent spaces between the words. Without interaction between the spoken (or written) word and silence, dialogue and communication become, if not completely impossible, at least much more difficult. Based on the proverb *speech is silver, but silence is golden*, I have previously highlighted the question of whether it is silence or the spoken word that is most highly valued in schools. This chapter will therefore consider more closely the importance of silence in school and other pedagogical settings, including the value of allowing silent pauses in anticipation of a response, silence as a pedagogical strategy and inner versus outer silence in today's students.

© The Author(s) 2020
E. Alerby, *Silence within and beyond Pedagogical Settings*,
https://doi.org/10.1007/978-3-030-51060-2_4

By reviewing the steering documents that govern schools in Sweden, it is easy to establish that the spoken word is often mentioned as an important part of the students' skills and learning process (Ministry of Education and Research, 1994; Swedish National Agency for Education, 2011). Presumably, the situation is the same in many other countries. Being able to talk is therefore an important skill that is valued highly in the school world, essentially regardless of the country in which the school is situated.

Amongst other things, the Swedish syllabus for Modern Languages clearly emphasises the spoken and written word. When teaching Modern Languages, schools should strive to help students develop their ability to 'use the language for communication in speech and writing,' and to 'share actively in conversation and written communication, express their own thoughts in the language, and grasp the opinions and experiences of others' (Swedish National Agency for Education, 2010a). Other subjects, too, emphasise the importance of students expressing themselves orally and in writing. One such example is Mathematics, where the syllabus states that one of the objectives is to strive to help students to be able to argue their point both orally and in writing (Swedish National Agency for Education, 2010b). The grading criteria for several different subjects show that both the spoken word and written word are to be assessed and graded. For example, in order to receive the Pass with Special Distinction grade in Modern Languages, students must, amongst other things, be able to actively participate in day-to-day language and be able to communicate in writing (Swedish National Agency for Education, 2010a).

What happens, then, if a student does not actively share in a discussion but is quiet, even though the point of discussion is such that they should have something to say about it? How do teachers and other students react towards a quiet student? Allow me to exemplify how silence can present itself in school by taking a look inside an ordinary classroom.

It is 8:15 on a Monday morning in the beginning of September, and a group of 12-year-old students is running along the corridor to be on time for the first lesson of the day. They are laughing and talking, arguing and shoving—they are thus far from silent as they make their way towards class. Once inside the classroom they continue to talk and laugh, even after the lesson has begun according to the timetable. The teacher at the front of the classroom does her utmost to get the class to 'calm down', as she puts it. When she has finally managed to get the students quiet, she begins the lesson by initiating a discussion on value issues. Some of the students are completely silent, not saying a word.

As described in this example, paradoxical situations often arise where, at one moment, a teacher is trying to get the students to quieten down, whilst in the next moment, students are expected to make their voices heard. On the one hand, it is not only during advanced and dangerous circus acts that silence is desirable—it is also sought after in various educational settings. On the other hand, teachers, just like circus performers, would like to see reactions and responses during discussions and question-and-answer sessions in the form of oral participation, in a similar way to how a circus performer would like to hear applause and cheers from their audience.

The Value of Waiting

In connection with the often paradoxical situation where the relationship between silence and talking in the classroom can shift back and forth, a question sometimes arises: How long does a teacher allow silence to continue after asking a question before personally filling the silence with the correct answer or asking another question? Loughran (1996), and originally Rowe (1974, 1986), terms this quiet moment *wait-time* and believes that teachers are often too quick to fill the silence that can occur between a question and its answer. Studies show that teachers allow on average just 0.9 seconds of silence after a student has responded to a question before the teacher themselves comment on the answer or ask another question (Rowe, 1974). Rowe also studied another type of 'wait-time': that between a teachers' question and when a student begins to answer. The conclusion Rowe came to is that the pause between a teacher's question and a student's answer is, on average, one second. It is important to be aware that this 'wait-time' consists of silence and to allow time for it in educational settings.

When teachers were made aware of the significance of the 'wait-time' between questions and responses, and were trained to increase the length of time they allowed for both types of 'wait-time' from about one second to an average of three seconds, a change was seen in the way students answered and reasoned on the questions. The length of the students' answers increased from an average of seven words per response to twenty-seven when the 'wait-time' was increased by two seconds. The number of contemplative and speculative answers also increased, and students began to compare and discuss their answers amongst themselves. Furthermore, the proportion of questions initiated by students increased by 25% and the percentage of incorrect answers was reduced (Rowe, 1974). Rowe states that, by allowing a longer 'wait-time' between a question and answer (even though an increase from one to three seconds may seem relatively small), teachers increased their expectations regarding the performance of weak students. By slowing down the speed of communication in the classroom and allowing quiet pauses, the quality of interaction in the classroom was increased. It is thus important to both *give* and *take* silence.

Giving and Taking Silence

It is not only in the classroom where we need to allow silent pauses when conversing with others—it is something that applies to all human communication (Scollon & Scollon, 1987). A conversation's quality increases when we allow silent moments and time for contemplation and reflection. The question is whether this is being observed and appreciated in today's society, where much depends on answers being given quickly without time for consideration. Almost daily, we are exposed by the likes of the media to various programmes and competitions, especially quizzes, that require answers to be given immediately. One example of this is Jeopardy,[1] where the answers are expected to be given with lightning reflexes, without any time for thought and reflection. Instead, it is speed that counts, although the answer given must, of course, be correct.

Another example where answers are expected immediately following the questions is political debates. If a politician were to delay in answering a question, staying silent for a moment, many people would see it as a sign of uncertainty, or possibly arrogance. In politics, though, there is a difference between a debate and a speech when it comes to how silence is used and valued. Unlike political debates, political speeches often contain pauses filled with silence. This is not in itself particularly noteworthy, considering that political speeches are usually well prepared and written by experienced speechwriters who use silence stylistically to emphasise and

---

[1] An American television game show that has been broadcast for many years and gained followers in many other countries around the world. The purpose of the competition is to find the right question to a provided answer as quickly as possible. The one who finds the correct question in the least amount of time wins.

highlight ideas and arguments, as well as to make an impression on the listeners and solicit their applause (Jaworski, 1993).

In connection to this, then, we can stop and think about what sort of question-and-answer culture prevails in school. Is it perhaps the case that the one who gives the fastest and loudest answer 'wins', so to speak? Within the classroom walls, there may be a culture where only the correct answers count, which causes some students to stay silent out of fear of answering incorrectly. To counteract this, both teachers and students should strive for a climate in which it is permissible to reflect and experiment with responses, to be able to give an answer that is perhaps not what was really requested when the question was asked. Perhaps this unexpected answer will be the one that leads thoughts to new dimensions that had not previously been thought of. So it is important to encourage openness and responsiveness towards others and what is said.

Silence as a Teaching Strategy

There is often a fluid boundary between the silence that is desired when, for example, a teacher is instructing or assigning tasks and the silence that is often used as a strategy to promote learning or maintain order in the class. There are probably many of us who recognise the well-known teacher's trick where the teacher themselves stays quiet and calm when students are not well behaved instead of trying to bring about silence by raising their voice.

Just like parents or the adult world in general, teachers often also use rhetorical questions, where an answer is not expected. 'What are you doing?' asks a teacher angrily when a student pushes one of their classmates in the school grounds. The teacher can very well see what the student is doing, and so continues the questioning, or perhaps reproof, without giving time for the student to answer; so the student stays quiet. For the teacher, who does not expect an answer, the student's silence is unlikely to be a problem. For the student, though, maybe the push was provoked (which in itself does not justify the push, but it does explain the reason). But, because the student had neither the opportunity to speak nor the ability or will to take that opportunity, they stay silent. If, on the other hand, the teacher asks, 'Why are you pushing?,' followed by a silent pause, then a response from the student is expected. If no answer is given in spite of the pause that follows the teacher's question, it may be perceived as provocative by the teacher whose question receives no response.

However, there could be many reasons for the student's silence—perhaps they feel ashamed, have no good answer to give or want to protect themselves or someone else.

Students can often be silent in subordination to their teachers, though this does not necessarily mean they are being submissive. A student's silence could be in protest to show their unwillingness to accept the teacher's decision or request (Jaworski, 1993). But this is not a complete description of the silence shown by students. As I understand it, the silence of a student is more multifaceted than that. A student can be silent for many reasons, such as shyness, fear, contemplation, demonstration, caution or a desire to find tranquillity.

INNER VERSUS OUTER SILENCE

Another aspect of silence that is worth noting in connection with educational situation is the distinction between inner and outer silence. Many students today are so boisterous and noisy inside that they fail to hear the silence even when it is there. The opposite is also true—it is possible to find inner silence in a noisy classroom, even though it is of course more difficult to create an inner calm in a noisy classroom than in quiet surroundings. Some people find it easier than others to shut out the external noise, so to speak. Whoever it may be, though, silence is an important part of learning, because it can provide opportunity for contemplation and reflection, amongst other things.

Peace and quiet in the classroom is essential, according to students themselves; but they also emphasise that it does not need to be completely silent. One student compares it to a scale of 1 to 10, where a quiet classroom would be 2. What is important is to cause as little noise as possible, which can be achieved by, for example, whispering and using small letters, as the students put it (Alerby & Kostenius, 2011; Bergmark & Kostenius, 2012).

Many people today are concerned about the steadily increasing noise levels in society. Perhaps our sound world is falling apart. Is silence being nibbled away at the corners through the exploitation of sound, with the risk of polluting our world? This is exactly what Mendes-Flohr (2012) emphasises—that people today are flooded with 'the cacophony of urban life' (p. 12). It is never quiet in today's society, and the same is true of today's schools.

* * *

It is valuable, and maybe even a necessity, to have knowledge and insights about silence, or as Jaworski (1993) expressed it: 'if we know more about silence, we will know more about ourselves' (p. 25). In this chapter, I have argued that silence is both essential and indispensable in various pedagogical settings. Silence can, for example, be used as a teaching strategy, where it is important to understand the value of allowing silent pauses and providing opportunity for giving and taking silence. It is relevant in this context to note that noise levels in society today are increasing. Some even believe that noise pollution is one of the greatest environmental problems facing society.

Regardless of whether or not our society is being polluted by noise, and silence is perhaps increasingly becoming seen as a scarcity, it may be of interest to ask children and young ones at school what they think about it. What do students think about silence or the lack of it in school? How do they value silence? In order to answer these questions, some students in a school were asked to use various linguistic and pictorial expressions to put their perspective into words, as is described in the next chapter and the second part of this book, 'Considerations of Silence in Day-to-Day Life at School: Some Experiences from Students.' I will also problematise and reflect on various forms of silent communication that can occur in surveys.

References

Aasland, D. G. (2012). Coming home to silence. In H. Fiskå Hägg & A. Kristiansen (Eds.), *Attending to silence. Educators and philosophers on the art of listening.* Kristiansand, Norway: Portal Academic.

Alerby, E., & Kostenius, C. (2011, July 27–30). *Silence for health and learning – A phenomenological reflection.* Paper presented at the 30:e International Human Science Research Conference, Oxford, UK.

Bergmark, U., & Kostenius, C. (2012). Student visual narratives giving voice to positive learning experiences – A contribution to educational reform. *Academic Leadership Journal, 10*(1), 1–17.

Jaworski, A. (1993). *The power of silence. Social and pragmatic perspectives.* London: Sage.

Loughran, J. (1996). *Developing reflective practice: Learning about teaching and learning through modelling.* London: Falmer Press.

Mendes-Flohr, P. (2012). Dialogical silence. In H. Fiskå Hägg & A. Kristiansen (Eds.), *Attending to silence. Educators and philosophers on the art of listening.* Kristiansand, Norway: Portal Academic.

Ministry of Education and Research. (1994). *Curriculum for the compulsory school system, pre-school class and the leisure-time centre—Lpo94.* Stockholm: Fritzes.

Rowe, M. B. (1974). Pausing phenomena: Influence on the quality of instruction. *Journal of Psycholinguistic Research, 3*(3), 203–224.

Rowe, M. B. (1986). Wait time: Slowing down may be a way of speeding up. *Journal of Teacher Education, 37,* 43–50.

Scollon, R., & Scollon, S. (1987). *Responsive communication.* Haines, AK: Black Current Press.

Swedish National Agency for Education. (2010a). *Kursplan för moderna språk* [Syllabus for modern languages]. www.skolverket.se/sb/d/2386/a/16138. 22 July 2010.

Swedish National Agency for Education. (2010b). *Kursplan för matematik* [Syllabus for Mathematics]. www.skolverket.se/sb/d/2386/a/16138. 22 July 2010.

Swedish National Agency for Education. (2011). *Läroplan för grundskolan, förskoleklassen och fritidshemmet: del 1 och 2* [Curriculum for the compulsory school system, pre-school class and the leisure-time centre: Part 1 and 2]. www.skolverket.se/content/1/c6/02/38/94/Lgr11_kap1_2.pdf. 1 Apr 2011.

Considerations of Silence in Day-to-Day Life at School: Some Experiences from Students

CHAPTER 5

The Value of a Quiet Place in School

Which part of the school, including the outside areas, is the students' favourite place? What do they appreciate about it? What do they do there? These questions come from a research study that focused on the way students experienced the spatial design of their school (Alerby, 2004). Students who participated in the study attended years 5 and 6 in primary school in Australia.

When students were asked to identify the most important place in the school—indoors or outdoors—they chose an area of the ground called the Peace Area. What was it about this place that was so important to the students? The students said that the Peace Area was important because it was a calm and quiet place in the school grounds. My intention with this chapter is to highlight and discuss how students experienced that quiet place, as well as their need for peace and quiet, or, as one of the students puts it, 'I like it because it's quiet and I can be alone and think and watch life go by.' By having an opportunity to sit by themselves at some point during the school day, this student is able to find an alternative to an otherwise intensive and stress-filled existence, to instead spend time in a place that allows for dialogue, learning, knowledge and participation (Stern, 2016).

Meanwhile, it is not only students who need calm and peaceful places at school; the same applies to teachers. Quiet places are needed for both students and teachers—for everyone at school, which Lees (2012) expresses as follows:

© The Author(s) 2020
E. Alerby, *Silence within and beyond Pedagogical Settings*,
https://doi.org/10.1007/978-3-030-51060-2_5

Given the intense pace of the school day, with shouted instructions, ringing bells, movement between activities, lessons, objectives, venues, multiple social interactions, it is no wonder that people in schools might want time out in silence. It is more surprising that not all schools appreciate this human need. (Lees, 2012, p. 101)

A study by Stern (2014) shows similar observations—that the participating children and adults expressed positive feelings towards, and a desire for, solitude in school. An opportunity to be by oneself in a quiet, calm and peaceful place during the school day is thus highly valued.

A calm, peaceful and quiet place is something the aforementioned school had in its grounds. Let us therefore begin by taking a closer look at the grounds belonging to this school. On the one hand, it is a regular class 1–6 school situated in one of Melbourne's suburbs in the state of Victoria, Australia. On the other hand, this school is quite unique in the architectural design of the school building itself as well as its grounds. Built and inaugurated in 1975, the school and its buildings consist of three different teaching facilities: a school library, public areas and an administrative unit. The teaching facilities contain classrooms that are completely open, where two classes meet together with their teachers in each facility. The entire school is built on a slope with school grounds that are proportionately large in comparison with the buildings and the number of students at the school. The playground contains an amphitheatre where all students and teachers gather each week, as well as two well-equipped play areas, with swings, climbing frames and slides. Additionally, the school grounds contain large open grass areas and a cricket pitch, as well as the quiet place— *the Peace Area* (see Fig. 5.1).

The Peace Area is situated in a corner of the school grounds, which, in turn, are in the immediate vicinity of the classroom for students in years 3–4 and 5–6. The area slopes a little and measures approximately 25 × 35 metres. It contains a few large trees that provide shade, as well as some smaller trees and shrubs. There are a number of wooden benches, some of which are grouped together, whilst others are placed on their own. The Peace Area is thus created by people for people, and the place could be compared with Cooper's (2008) exploration of the significance of the garden. Cooper says that the garden is like a revelation of a close interaction between people's creative activities in the world and 'the mystery' of what it is that actually makes the world exist for them. Although this quiet place

Fig. 5.1 The Peace Area. (Photo by Eva Alerby)

in the school grounds is not quite like a garden, it is created by means of human 'creative activities' in a world that exists just for them.

The vegetation, together with the peace and quiet that prevails in the Peace Area, is most likely part of the reason so many birds visit. So the idea is that it should be *quiet* here. Students are not allowed to be noisy or engage in loud activities in this place. Instead, students who spend time here should be given an opportunity for peace and stillness. If they would like to join in noisy activities, such as laughing, shouting or perhaps ball games, students can go to other areas of the school grounds. But how do the students perceive this place? Why do they identify this area as the most important place in school?

Voices from a School Yard

When we are trying to find out the way students perceive things and so on, their voices are often expressed when we talk with or interview them. But since language consists of so much more than verbal communication (or written, for that matter), students in this case were asked to portray their experiences by means of pictures. The pictures that are of interest are the students' drawings and photographs.

Students were given the task of drawing the most important place at school, whether inside the building or somewhere in the school grounds—the choice was theirs. They were also allowed to photograph three important indoor and outdoor places. Based on their drawings and photographs, the students then talked about the places they had portrayed and how they felt about them. When the drawings, photographs and verbal comments were analysed, it could be seen that more than half of the students had depicted the Peace Area in their drawings (see Fig. 5.2), and all of them had photographed the same place as one of three important outside areas. Why is this place so important for the students? What do they do there?

Fig. 5.2   The Peace Area depicted in a drawing by one of the students

When the students talked about their drawings and photographs, it was found that they enjoyed and appreciated the Peace Area for the very reason that it is a calm and quiet place where they can find stillness. It is a place where they can sit calmly in peace and quiet to relax, think and reflect by themselves, read and work on their school assignments, and also to chat quietly with their friends. One of the students expressed this as follows when talking about their drawing and experiences about the Peace Area:

> The Peace Area is my favourite place because it is quiet. It is important just because it is quiet. I get peace and quiet and I rest here or I finish my work. I can get away from the rush of people when I'm here. It's a very relaxing place and I really enjoy being here. I often use it at play time and lunch time to read, draw, watch birds or sit and relax.

Another student says this about the same place:

> It's very quiet and it's a place where we can do quiet stuff ... or you can go there and breathe. ... When you are tired, you can go there and be for yourself. It's nice and quiet and you can talk to friends and watch the birds and watch nature. And no one makes any noise here. ... It's good, I like it. I like it because we can just sit there and talk and dream and so on.

The students clearly express a need for being able to retreat to a quiet and calm location to find peace and stillness. They can spend time here in peace and quiet, and, according to the students, the place makes it easier to reflect and contemplate during the school day.

Everyday life and activities at school must be viewed as especially active rather than contemplative. In the active life of school, people meet and participate in various forms of social activity, which is the type of life referred to as *Vita Activa* by Arendt (1958). Arendt emphasises that humans are by nature active and social creatures that share in social activities. Active life *(Vita Activa)* includes the elements of labour, work and action. In contrast to this active life is the human need of withdrawing to find peace and stillness, which is where the contemplative life *(Vita Contemplativa)* takes over, which—according to Arendt—consists of the elements of reasoning, knowing and thinking.

How, then, should we understand the significance of a quiet and tranquil place in the school world? Silence can help a person to listen to their inner self as well as to others, which the students express clearly. They also stress the need of being able to spend time in a place where they can relax together with others or just sit with their own thoughts for company, or as one of the students put it: 'you can go there and breathe.'

THE NEED FOR A QUIET AND PEACEFUL PLACE

The school yard should be seen as an important and meaningful place, and according to some students, it is during the breaks, when they use the school grounds, that they encounter positive experiences (Alerby, 2003). Previous research shows that students perceive the school environment as an expression of how the adult world values them (see, e.g., Skantze, 1989), and the same is likely true of the grounds—the school grounds can also be considered an expression of how students are valued by the adult world. According to this reasoning, we might wonder what values the school grounds express. How do students who spend time there perceive the way the adult world values and cares about them?

The vast majority of school grounds consist of paved areas instead of grass, trees or other vegetation. Most do not have particularly attractive areas for students to play in, and, if there is a play area, it is not usually very well equipped. But this is not true of the school we have been considering in Australia. At this school yard, students have access to spacious areas, and the Peace Area is not the only place where students can spend time amongst vegetation. Instead, it is the silence and stillness of the place that are important and meaningful to the students. Silence and stillness needs to be viewed as an essential and inevitable part of life (Alerby & Elídóttir, 2003). Mother Teresa says that we humans need silence to touch each other's souls. It is that ability to touch another's soul that should be of great importance in different pedagogical settings, regardless of whether the person in question is a student, a teacher, a principal or a janitor.

The students who have expressed themselves in this chapter clearly state that the Peace Area is a place to go to in order to 'get away from the rush of people' and instead spend time in silence and stillness. Silence is, however, something more than 'just the absence of speech or language' (van Manen, 1990, p. 112). Studies show that we humans generally find noise to be stressing. It can once again be emphasised that high sound levels affect our bodies, resulting in high blood pressure, an increased heart rate and the release of extra stress hormones (Babisch, Fromme, Beyer, & Ising, 2001; Englund, 2000). These high sound levels are going to be our greatest public health issue in the twenty-first century, and sound pollution is a major problem in today's society, according to Passchier-Vermeer and Passchier (2000). What we might stop to consider in this context is how sound levels at our schools contribute to the stress of those who spend time there. These students clearly expressed the importance of spending time in silence during the school day: 'The Peace Area is my favourite place because it is quiet. It is important just because it is quiet.'

Students spend several years in school, and these years are most likely going to have an impact on their future lives in one way or another. How students perceive school is therefore important, not only during the time they spend in school, but also for their future lives. The study described in this chapter shows clearly how students need to be able to withdraw into a place of peace and stillness—of silence. This opportunity is likely to leave an impression on their perceptions and memories of their school years, which one of the students mentions when she says: 'This place [the Peace Area] is important to me because it is where I have spent time over seven years. I've got lots of memories from my time spent under the wattle tree.'

*   *   *

This is an example of a study in which students themselves have highlighted the value of spending time in silence during the school day. In the next example, I will highlight and discuss silence from a completely different perspective; namely, the existence of silent communication in surveys, and how it can be understood.

REFERENCES

Alerby, E. (2003). 'During the break we have fun.' A study concerning pupils' experience of school. *Educational Research, 45*(1), 17–28.

Alerby, E. (2004). The appreciation of a quiet place at school. *Didaktisk tidskrift, 14*(1), 57–61.

Alerby, E., & Elídóttir, J. (2003). The sounds of silence: Some remarks on the value of silence in the process of reflection in relation to teaching and learning. *Reflective Practice, 4*(1), 41–51.

Arendt, H. (1958). *The human condition.* Chicago: Chicago University Press.

Babisch, W., Fromme, H., Beyer, A., & Ising, H. (2001). Increased catecholamine levels in urine in subjects exposed to road traffic noise: The role of stress hormones in noise research. *Environment International, 26*(7–8), 475–481.

Cooper, D. E. (2008). *The philosophy of gardens.* Oxford: Oxford University Press.

Englund, A. (2000). *Trafikstress. En redovisning av utförande och resultat i KFB-finansierade forskningsprojekt 1996–1999* [Traffic Stress. A description of the performance and results in the KFB funded research 1996–1999]. Stockholm: Fritzes.

Lees, H. (2012). *Silence in schools.* London: Trentham Books.

Passchier-Vermeer, W., & Passchier, W. F. (2000). Noise exposure and public health. *Environmental Health Perspectives, 108*(1), 123–131.

Skantze, A. (1989). *Vad betyder skolhuset? Skolans fysiska miljö ur elevernas perspektiv studerad i relation till barns och ungdomars utvecklingsuppgifter* [What is the significance of the school building: The physical environment of the school seen from a pupil perspective and studied in relation to their developmental tasks]. Doctoral thesis, Stockholm University, Stockholm.

Stern, J. (2014). *Loneliness and solitude in education: How to value individuality and create an Enstatic school.* Oxford: Peter Lang.

Stern, J. (2016). Solitude and spirituality in schooling: The alternative at the heart of the school. In H.E. Lees, & N. Noddings (Eds.), *The Palgrave international handbook of alternative education* (pp. 431-445). London: Palgrave Macmillan. Chapter 28.

van Manen, M. (1990). In State University of New York Press (Ed.), *Researching lived experience. Human science for an action sensitive pedagogy.* London.

CHAPTER 6

Silent Communication: Or 'Taxi Hell'

A great many people are likely to have been asked to respond to a questionnaire or survey at some point. Using surveys to find answers is quite common when, for example, organisations or authorities would like to have information from a large number of people, and even children and adolescents are expected to respond to various types of survey. Surveys that students are asked to respond to are often about their experiences at, or their view of, school, but can also involve how they are feeling—their state of health. It is not uncommon for these surveys to be distributed at school.

Various types of questionnaire or survey are also used in research, especially when the researcher would like to collect data from a large number of people (see, e.g., Eggeby & Söderberg, 1999; Wiersma & Jurs, 2005). The objective of surveys is, of course, to receive as good information as possible about the subject being asked, but the question is whether this is always the case.

Most people who have come in contact with a survey have likely experienced that it is not always easy to answer a certain question or statement by ticking a box with a fixed-choice answer. Their own answer does not always match the pre-printed alternatives. Different people handle this in different ways. Some simply leave the question unanswered, and some add their own response or write comments in the margin of the survey sheet. Others might pick an answer at random just so that something is selected,

© The Author(s) 2020
E. Alerby, *Silence within and beyond Pedagogical Settings*,
https://doi.org/10.1007/978-3-030-51060-2_6

and still others might deliberately choose to answer incorrectly, thus ruining the study's reliability.

One problem that arises in connection with this is that an unanswered question can be interpreted in different ways, if it is interpreted at all. Another problem is that any extra information in the form of additional options added in a multiple-choice question or comments written in the margin are often not considered valid responses when a survey is fed into a data processing programme.[1] Instead, any such extra information is left unprocessed. This, in turn, means that whatever was written is silent, or is silenced.

In this chapter, I therefore intend to discuss how unanswered questions and notes in the margin of a survey sheet might be understood—that is, making sense of silent communication. I will also reason on why a survey does not get completed in the way intended when it was designed and distributed.[2]

An unanswered survey question could be considered a non-message, to use Bateson's (1987) terminology. It might also be the case that when we cannot, or do not wish to, respond to a question or statement, we instead become silent. Polanyi (1969) believes that all people possess silent and unspoken dimensions within themselves, which is demonstrated in those situations when we lack the words to express ourselves—we become silent. That silence can, however, express many different messages, but these messages are easily ignored or disregarded. Regardless of the reason for silence, a non-message is still a message—the silence tells us something (Bateson, 1987). So, can an unanswered question, a mark by the side of a box instead of inside it, or some text written in the margin of a survey sheet be considered a non-message? And what does this non-message—or perhaps even better, this silence—communicate, in that case?

[1] The data processing programme Statistical Package for the Social Sciences (SPSS) was used in the study referred to in this chapter.
[2] This chapter is based in part on Alerby & Kostenius-Foster (2005) and Alerby & Kostenius (2011), and the content has been adapted and rewritten to fit with the theme of this book.

WHEN COMMUNICATION IN SURVEYS IS SILENT

'Taxi hell,' wrote a 12-year-old boy in the margin of a survey he was asked to fill out. These words form the basis of this discussion on non-messages in surveys, which could be considered silent communication. More about this later.

In order to better understand silent communication in surveys, a number of questionnaires that contained unanswered questions and written comments or messages in the margins were analysed. The surveys in question had been completed as part of Arctic Children, a research and development project[3] relating to psychosocial health amongst children and adolescents in the Arctic. It was done with the help of the WHO's standardised and commonly used Cross National Survey on Health Behaviour questionnaire (Swedish National Institute of Public Health, 2002). This questionnaire was developed in the early 1980s, and has since then been used to collect material on health indications and health behaviour amongst students in Europe, Canada and the United States (see, e.g., King et al., 1996; Williams & Currie, 2000). The version of the survey used in this study contained 74 questions, each of which contained a number of different answers to choose between. Four-hundred and forty students participated, and the response rate must be considered high, with over 90% responding, or 400 out of 440 surveys. This was likely due to the participants being allowed to use lesson time to complete the forms, which were thereafter collected by each class teacher.

When entering the data into the SPSS software, we discovered that many of the surveys contained several unanswered questions, as well as marks outside of checkboxes and comments written in the margins. All in all, just over 50% of the questionnaires that were handed in, or 208 of 400, contained unanswered questions and/or notes in the margins. This caught our attention and got us thinking about whether the students were trying to tell us something beyond what was possible using the pre-printed alternatives, but especially about how these unanswered questions and notes in the margins of a questionnaire can be understood. This has previously been referred to as silent communication, and we can now consider a few examples of what this could be in the form of *silent* and *silenced* messages.

---

[3] Arctic Children I—Development and Research Project of Psychosocial Well-being of Children and Youth in the Arctic, EU, Interreg. Kolartic 3A, 2003–2006 (see, e.g., Ahonen, 2010; Ahonen et al., 2008).

SILENT MESSAGES

Some students left entire questions or parts of them unanswered, as will be shown later.[4] What we might consider in this context is whether these students made active and conscious decisions to not respond to certain questions or statements. But whatever the case may be, there are no responses provided for some of the questions and statements, so the students' voices could be viewed as silent.

Silence is constitutive of discourse (Heidegger, 1992), and it is also important to be aware that, in order to be silent, a person must also have something to say. A lack of voice is, as mentioned previously, not the same as being silent. A person can only be considered silent when he or she can speak—that is, has something to say (Merleau-Ponty, 1996). So, do these students who are supposed to fill in a questionnaire have something to say about what is being asked, or not? But perhaps it is just as important to consider whether anyone is going to listen to what they are saying. Allow me to return to this shortly.

A person filling in a questionnaire might willingly refrain from answering one or more of the questions—that is, they can be silent; but, for another person, their silence may be forced due to being unable, or not daring, to express what is asked of them. In spite of this, the lack of an answer—the silence—conveys something. At this point, I could yet again state that silence becomes a language when words are insufficient or when a person prefers silence over words.

In the case of unanswered questions or statements, the student is expressing a non-message. These non-messages convey a message that might be just as important as messages that are spoken aloud, or, as in this case, written down in the form of a marked box or a written answer or message to the side of a question. Within the scope of this chapter's line of reasoning, however, the question is whether these non-messages, this silence, are taken into account. Another question is *why* a student might choose to not respond to a question, or part of one. The reason may be that the student is shy or unsure, or it could be a form of self-protection, where the student does not wish to openly confirm their answer to the question. It could also be due to the student deliberately choosing to thwart the study, so to speak, by not giving an answer. The example shown

---

[4] The examples used, the questions from the questionnaire and the students' written words in the margins have been translated from Swedish into English for the purpose of this book.

in Fig. 6.1 may be a case of a student not wanting to openly respond to some of the statements provided.

In the question shown in Fig. 6.1, it is easy to establish that the statements left without a response are the ones to do with psychosocial aspects such as fear, nervousness and problems sleeping, whilst statements focusing on physical complaints have been answered. A little over half of the question's statements are silent, but for what reason it is impossible to say. Another example of a question that can be perceived as sensitive is the one shown below (Fig. 6.2).

20. How often, during the last 6 month, have you had following problems? *(Put an X at each row)*					
	Almost every day	More then once a week	Ap. once a week	Ap. once a month	Seldom or never
	1	2	3	4	5
Head ache	☒	☐	☐	☐	☐
Stomach ache	☐	☐	☐	☒	☐
Back pain	☐	☐	☒	☐	☐
Felt down	☐	☐	☐	☐	☐
Irritated or in a bad mood	☐	☐	☐	☐	☐
Felt nervous	☐	☐	☐	☐	☐
Sleeping problems	☐	☐	☐	☐	☐
Felt dizzy	☐	☐	☐	☐	☐
Neck- and shoulder pain	☒	☐	☐	☐	☐
Felt scared	☐	☐	☐	☐	☐
Tired and exhausted	☐	☒	☐	☐	☐

Fig. 6.1 An example where parts of a question have not been answered—where parts of the message are silent

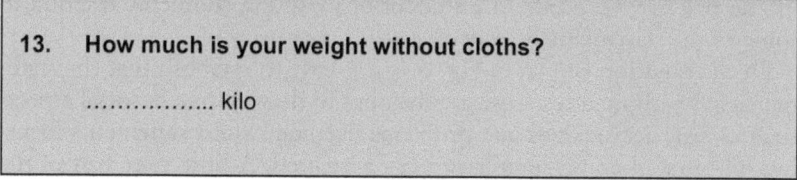

13.   How much is your weight without cloths?

.... kilo

Fig. 6.2 The question 'How much do you weigh without clothes?' was left unanswered by several of the participating students

Several of the students left the question about their body weight unanswered, which might be because they do not wish to talk openly about how much they weigh. Students that Kostenius (2008) has encountered in her research emphasise that trust affects the degree of openness. How much could the students trust the ones who were going to have access to the information in the survey, and how would the information be used? Another possible reason why the question might not have been answered is that the student might simply not know the answer. Regardless of the reason, the question becomes silent, since the answer is missing.

The example questions shown in Figs. 6.1 and 6.2 may be perceived as too personal, and in an attempt to try and protect themselves from being open about the content of the question, some of the participating students chose not to respond to the questions and statements—they are silent, instead.

A person can be silent for many reasons (Bollnow, 1982). Silence can be used as a form of self-protection and can, for example, be used to shut oneself off from the outside world—a way of protecting oneself, someone else or something that the person knows about but does not want to talk about. A person might choose to answer a survey question for many different reasons, such as because of social expectations or the nature of the situation. Other, more fleeting, factors can also have an impact on how a person responds, such as their current mood, external circumstances in the form of what is happening around the world at the moment or perhaps even seasonal variations such as the weather (Schwarz & Strack, 1991, 1999). All of these various factors can also affect the non-answer—the silent message.

The reason why a person chooses not to answer a survey question is probably a combination of several different causes. However, a non-message must be considered to be a silent one—a message that can convey something, even though the student's voice is silent.

SILENCED MESSAGES

Some students put a mark between or to the side of checkboxes (see Fig. 6.3), which signals that the pre-printed responses did not match the students' own answers.

In the example shown above, the student has drawn three of their own extra boxes (6, 7 and 8) in order to really emphasise that they absolutely do not look forward to going to school. The problem, though, is that this is not counted as a valid answer to feed into the software that will analyse the responses, so the information that the student wishes to convey is not taken into account. Instead, it could be said that the student's voice is silenced.

Another example of the printed answers not matching a student's own response, or where a student wishes to explain their answer, is the quote in the introduction to this chapter where a boy wrote 'Taxi hell' under one of the questions (see Fig. 6.4). The question was: 'How many days per week do you usually spend time together with friends directly after school?,' and the possible answers were: 0, 1, 2, 3, 4 or 5 days per week.

65.	In the following there are some more statements about school *(Put an X at each row)*				
		I totally agree	I agree	I'm doubt-ful	I'm not agree	I'm absolutely not agree
		1	2	3	4	5 6 7 8
	1. I look forward to school	☐	☐	☐	☐	☐ ☐ ☐ ☒
	2. I like to be in school	☐	☐	☐	☒	☐
	3. It is a lot in school which I don't like	☐	☐	☒	☐	☐
	4. I wish I didn't have to go to school	☐	☐	☐	☒	☐
	5. I like what we do in school	☐	☐	☒	☐	☐

Fig. 6.3 An example of extra information being supplied in the form of self-constructed responses in the margin of a questionnaire

Fig. 6.4 An example of extra information that has been written below the question

This boy chose the box for one day per week and also wrote 'Taxi hell' in the margin, because he had to ride the school taxi to and from school every day, and the taxi only picked him up and dropped him off right at the beginning and end of the school day. In this boy's municipality, students who ride the school taxi were not allowed to bring a friend home with them in the taxi, either. This made it essentially impossible for him to socialise with friends after school.

The way the question was formulated, however, did not allow the boy to explain why he was only together with friends after school relatively rarely. Without this written comment, we might have speculated on a number of possible reasons, such as: Maybe he does not have any friends? Perhaps he does not want to be together with friends after school? Maybe he is not allowed to take friends home with him? And maybe he also is not allowed to go to a friend's home himself? The main argument here is that, in a survey such as this, we can never know the reason for a provided answer unless the student themselves writes something in the margin. A message written in the margin of a questionnaire will, however, almost certainly be silenced—it will not be considered.

Schütz's (1999) interpretation of the everyday life-world and his attempt to structure social life warrant our consideration in this context. Schütz divides the life-world into a *world of directly experienced social reality*, a *world of contemporaries*, a *world of predecessors* and a *world of successors*. The 'world of directly experienced social reality' is in turn separated into a *thou-orientation* and a *we-relation*. In a 'thou-orientation,' a person is unilaterally aware of another person's lived existence but does not have an active part in that person's life. If, instead, a person has a 'we-relation' with another individual, then the two are mutually aware of each other and share in each other's lives. We might consider whether the questions in a survey are constructed according to a 'thou-orientation' or a 'we-relation.' It is quite easy to confirm that those who design a survey are very unlikely to be actively participating in the lives of the people who are asked to answer the survey questions. The survey designers are instead only unilaterally aware of the lived experience of these prospective participants.

In the examples I have given, students in some cases have tried to convey something more than the questionnaire allows for or asks for. But the question is whether these messages are taken into consideration; that is, is anyone listening to what the students are conveying? Sometimes, a person, like the boy in this example, wishes to provide an explanation for their response, which the boy did by complementing his selected checkbox with additional words. But these were not listened to—they were not heard when the data was later analysed. His message was not taken into consideration—it became silent.

Another question is why a student might not answer a question at all. Under the *Silent messages* heading, it was reasoned that one possible reason could be shyness or tentativeness. Another reason could be that the student is uninterested in the survey or tired of answering the questions. It is clear that some students get bored when completing a questionnaire, especially if it contains many detailed questions. One of the participating students expresses as much in Fig. 6.5.

'I'm tired of the schoolwork,' and the box for 'I totally agree' is selected, followed by the comment: 'and of this questionnaire!' However, any extra information that is written in questionnaires is in danger of being excluded from the analysis, instead becoming *silenced*.

A Silent Message Is Also a Message

A survey must be regarded as limited to the extent that it does not allow for any information to be provided other than exactly what is requested, which means that some details are left unsaid. Silence thus takes on different forms when a questionnaire is filled in—the one who completes the

62.	In the following there are some statements concerning experiences of work at school (think about work at school as well as home work) (Mark each statement with an X.)				
		I totally agree	I agree	I'm doubt-ful	I'm not agree	I'm absolutely not agree
		1	2	3	4	5
	1. I have too much school work	☐	☒	☐	☐	☐
	2. I think the school work is hard	☐	☐	☐	☒	☐
	3. I get tired of the school work	☒	☐	☐	☐	☐
	and of this questionnaire					

Fig. 6.5 An example of one student expressing how boring it was to fill out the questionnaire

questionnaire can in some sense become silenced, or the message itself is silent.

Silence can take the form of a very heavy mark next to a checkbox, which can indicate anger or frustration. Silence can also be a checkbox that is left empty in silent protest, or as a sign of disinterest or lack of trust. This reasoning might lead to the interpretation or conclusion that written silence becomes a language in which the silence is used as a way of expressing something that is not requested or accepted, or when a person has not been given opportunity to express what they really want to say. There is thus something beyond what is said, something that cannot be communicated verbally—an implicit and silent language. At this point, we may again be reminded that Merleau-Ponty (1995) argues that a lack of voice is not the same as being silent. In order to be silent, a person must have something to say; that is, a person is silent only when they are able to speak and also have something to convey. Do students that complete a questionnaire have something to say, then? Are they silent because they are shy, afraid or perhaps even bored?

Regardless of the reason for the silence, it can be established that silent communication in surveys is something that exists, and it has been argued above that information expressed in a survey is sometimes either silent or silenced. In any case, silence does communicate something, as long as we are prepared to listen to it and think about what it means. And, regardless of the reasons for it—shyness, personal integrity, lack of trust, disinterest or boredom—there is value in silence. *A silent message is also a message.*

REFERENCES

Ahonen, A. (2010). *Psychosocial well-being of schoolchildren in the Barents region. A comparison from the northern parts of Norway, Sweden and Finland and Northwest Russia.* Doctoral thesis, University of Lapland Press, Rovaniemi.

Ahonen, A., Alerby, E., Johansen, O.-M., Rajala, R., Ryzhkova, I., Sohlman, E., & Villanen, H. (red.). (2008). *Crystals of schoolchildren's well-being. Cross-border training material for promoting psychosocial well-being through school education.* Rovaniemi, Finland: University of Lapland Press.

Alerby, E., & Kostenius, C. (2011). 'Dammed Taxi Cab' – How silent communication in questionnaires can be understood and used to give voice to children's experiences. *International Journal of Research & Method in Education, 34*(2), 1–14.

Alerby, E., & Kostenius-Foster, C. (2005, March 10–12). *A silent message is also a message.* Paper presented at NERA:s Congress, Oslo, Norge.

Bateson, G. (1987). *Steps to ecology of mind.* Northvale, NJ: Jason Aronson Inc.

Bollnow, O. F. (1982). On silence – Findings of philosophicopedagogical anthropology. *Universitas, 24*(1), 41–47.

Eggeby, E., & Söderberg, J. (1999). *Kvantitativa metoder: för samhällsvetare och humanister* [Quantitative methods: For social sciences and humanities]. Lund, Sweden: Studentlitteratur.

Heidegger, M. (1992). *History of the concept of time.* Bloomington, IN: Indiana University Press.

King, A., Wold, B., Tudor-Smith, C., & Harel, Y. (1996). *The health of youth. A WHO cross national survey* (European series 69). WHO Regional Publications.

Kostenius, C. (2008). *Giving voice and space to children in health promotion.* Doctoral thesis, Luleå University of Technology, Luleå.

Merleau-Ponty, M. (1995). *Signs.* Evanston, IL: Northwestern University Press.

Merleau-Ponty, M. (1996). *Phenomenology of perceptions.* London: Routledge.

Polanyi, M. (1969). *Knowing and being.* Chicago: University of Chicago Press.

Schütz, A. (1999). *Den sociala världens fenomenologi* [The phenomenology of the social world]. Göteborg, Sweden: Daidalos.

Schwarz, N., & Strack, F. (1991). Context effects in attitude surveys: Applying cognitive theory to social research. In W. Stroebe & M. Hewstone (red.), *European Review of Social Psychology* (Vol. 2, pp. 31–50).

Schwarz, N., & Strack, F. (1999). Reports of subjective well-being: Judgmental processes and their methodological implications. In D. Kahneman, E. Diener, & N. Schwarz (red.), *Well-being: The foundations of hedonic psychology.* New York: Russell Sage Foundation.

Swedish National Institute of Public Health. (2002). *Skolbarns hälsovanor 2001/2002* [School children's health habits 2001/2002]. Report number R 2003:50. Stockholm: Swedish National Institute of Public Health.

Wiersma, W., & Jurs, S. G. (2005). *Research methods in education: An introduction.* Boston: Pearson.

Williams, J., & Currie, C. (2000). Self-esteem and physical development in early adolescence: Pubertal timing and body image. *Journal of Early Adolescence, 20*(2), 129–149.

PART III

The Art of Appreciating Silence

CHAPTER 7

Closing Thoughts on Silence and Pedagogy

In this book, *Silence Within and Beyond Pedagogical Settings*, I have high-lighted and discussed some of the different aspects of silence, as well as various nonverbal modes of expression in which silence often plays an essential role. My argument is that silence is meaningful in many different pedagogical settings, as is the art of listening. The value of not only hearing what a person says, but really listening to them, cannot be overstated; and to do that, silence is needed.

Some students were given the opportunity to voice what silence means to them, such as the example of 'the Peace Area,' and, in the case of 'Taxi hell,' silent or silenced messages in questionnaires have been highlighted. And now, it is finally time to try and tie all of this together with a few clos-ing words—words that connect with what is already written. Amongst other things, these words will consider how the view of silence can change over time, what the level of acceptance might be for silence in today's society—especially in today's schools—and whether nonverbal modes of expression can speak, or whether they should be considered silent. I will also discuss the aspects of whether (and in that case, how) silence is valued in schools, and what is actually said in that which is not said. Finally, I will reflect on the question that was posed in the introduction to this book: Is it possible to talk (and write) about silence? But first, it is time to bring this book to a conclusion.

© The Author(s) 2020
E. Alerby, *Silence within and beyond Pedagogical Settings*,
https://doi.org/10.1007/978-3-030-51060-2_7

SILENCE THEN AND NOW

The fact that the absence of a message—a so-called non-message—is a message, that it conveys something, has been established by Bateson (1987). One example he gave of this is when someone sends a letter that never receives a reply. The lack of a reply—that is to say, the silence—conveys a message, and can be viewed as a language from that perspective. The question is how the silence is interpreted. There are several possible interpretations to this silence, such as forgetfulness, carelessness, disinterest, a conscious choice, perhaps in the form of rejection or an attempt to protect someone or something (or maybe it got lost in the post?).

In today's society, letters have been largely replaced with email, but the nature of silence is also found in the modern digitalised world. An email we send, which we expect a quick reply to, might never be answered—it is silent, instead. In this situation, the length of time between the message being sent and an expected reply is important. We often expect, or at least hope for, a quick response to an email, preferably by return of email. Although there is a difference between a letter sent through the regular mail service and an email sent digitally through cyberspace, the silence speaks its (un)clear language, regardless of what form the message takes. Although the form differs (a letter is sent through the regular mail service and an email is sent on its way with the click of a mouse) any silence that arises (when we do not receive an answer) speaks to us, even though it is not possible to know for sure what the reason is for a lack of response.

Another angle on the lack of response is related to how individuals and groups use social media, perhaps social network sites in particular (Boyd & Ellison, 2008). Facebook can serve as an example in this context, in which the lack of active participation on the part of certain individuals—their silence, if you will—can be understood and interpreted in many different ways. Boyd (2007) terms these individuals as an *invisible audience*; that is, they are there, they are able to observe and see other people's posts, but they do not make themselves heard—they are instead considered to be an invisible, or silent, audience.

In the school world, silence has been valued and used differently through the ages, although considerable variations can, of course, occur within a particular period of time. In general, it can be said that silence on the part of students was emphasised differently in the past to how is most likely the case in the schools of today. Historically, students were expected to be silent and only listen, whilst the teacher did the talking (Adelmann,

2008). Hedquist (2006) stresses that, when he went to school in the mid-1950s, he was 'firmly drilled to listen in an authoritarian school' (p. 24). Students were allowed to speak only when giving a direct answer to a question posed by the teacher.

Today, too, teachers expect students to be quiet and listen, and not disturb others or make noise during the lessons. However, students are usually given more leeway to talk nowadays, and are also expected to show initiative in discussions or other oral activities. The pattern of communication between teacher and student has shifted from vertical to a more horizontal and democratic one (Adelmann, 2002, 2008).

Listening is a communication skill that schools are expected to teach to their students. One school subject in which listening is both an important and unavoidable component is music; other subjects are languages. Listening can, however, differ, depending on whether the subject itself requires listening—such as being able to differentiate tones or nuances in pronunciation—or whether it is something a teacher strives for so that students can most effectively benefit from the material. It is worth noting that listening is a part of silence, and vice versa, and also that our attitude towards both listening and silence changes over time.

It is not only the attitude towards silence that has changed; the way noise is viewed has, too. Many people are concerned about the ever-increasing noise levels in society, which can result in our sound world falling apart. The high, sometimes even dangerous, noise levels that can occur in society, and especially schools, are often talked about (Swedish Work Environment Authority, 2013; Babisch, Fromme, Beyer, & Ising, 2001; Englund, 2000). But it was as early as the eighteenth century that we realised that noise can be a problem. One reason for this was that many of the major European cities were overpopulated, which led to a 'thickening of the carpet of sound' (Englund, 2005, p. 30). But the fact that noise levels in society are becoming louder is also resulting in silence being re-evaluated, appreciated and refined. The landscape of sound, as well as silence, can thus be said to have changed over time. What was considered acceptable sound and acceptable silence in, for example, the eighteenth century, is unlikely to be the same in our time.

An indication that silence is valued and sought after in today's society is that some environments have been transformed into *quiet areas*. One concrete example of this that was previously highlighted is the quiet place set aside in the grounds of a school in Australia: the Peace Area. Another example is that some train compartments or carriages have been

designated as quiet zones. In these special, quiet, compartments and carriages, it is, for example, not permitted to talk on a mobile phone. Although
these compartments and carriages are marked as quiet, it is not uncommon for travellers to 'have music plugged directly into their ears' (Maitland,
2009, p. 3). Other examples of quiet areas are found in restaurants that do
not allow guests to have their mobile phones switched on. In these examples, silence has been reduced to the absence of the ringtones of mobile
phones, together with the words spoken in connection with a mobile-
phone call. That is, silence is in these cases 'the absence, not of sound per
se, but of noise which is obtrusive or salient' (Cooper, 2012, p. 55).

Attitudes towards silence and noise are thus changing over time as society develops. How much silence are we really inclined to accept today?
And who decides what silence is okay and what is not?

Permissible Silence?

At the beginning of this book, I asked who it is that sets the rules for what
kind of silence is, or is not, acceptable. It is surely difficult to find a defini-
tive answer to that question. The way that silence is viewed differently
across cultural boundaries was highlighted earlier, where different cultural
groups can view silence and handle it in different ways. These cultural
groupings may consist of different nationalities, but the way silence is
viewed can vary even within the same country. Additionally, a group of
people consists of different individuals who can each view and understand
silence in different ways.

Considering the situation at school, there are different school cultures,
within which silence is viewed and treated in different ways. Something
that these all have in common is that the spoken word is often put ahead
of silence. In light of this, we must ask why a quiet person—perhaps a
school student—is often perceived as provocative. Could it be that activity
and verbal skills are preferred to contemplation, stillness and silence in
today's Western society—in various pedagogical settings in particular?

If so, this is something that could be called into question. Most people
have a need to occasionally take a step back to think, and this is where a
quiet and tranquil place can help a person to 'make sense of thoughts,
emotions, actions and the context in which these are embedded' (Alerby
& Elídóttir, 2003, p. 42). On this basis, silence can be viewed as a sense-
making process.

Studies by von Wright (2012), however, have shown that quiet students are sometimes viewed by the school, teachers and principals as a problem, and that a student's silence can even be described as pathological. Von Wright instead describes students' silence in more rational terms, more specifically through the asymmetrical relationship between teacher and student: '[A] vital component of an educational relation is *asymmetry;* the asymmetry between a teacher who has the authority and young students who are on their way to entering public life' (von Wright, 2012, p. 93). In this educational relationship, there are always facets of silence that are comprised of both expectations and uncertainty.

When it comes to silence in different pedagogical settings, it is important to be observant as to how the silence is perceived—negatively or positively. In some contexts, silence is needed in order for reflection and learning to occur, whilst in other situations, it can be completely devastating. If we use discernment to allow silence, we are preparing the way for those who are not so talkative. We need to be aware that a reticent person is not automatically silent in thought or mind. In pedagogical settings, we should instead try to draw out the silent knowledge and silent languages that, according to Polanyi and Merleau-Ponty, as well as others, can be found in everyone—all to promote learning. One condition for being able to succeed in these efforts is to value (the good) silence and allow opportunity for perceptions and experiences to be expressed and communicated with the help of as many languages as possible—both verbal and nonverbal.

To Speak Or Not to Speak: That Is the Question

The meaning of silence could be described as a condition without sound, and it is not difficult to establish that the written word or a picture do not generate any particular amount of auditory stimulation. The written word and images are both silent by nature. Could it be that symbolic systems—nonverbal modes of expression—can speak? If they can, are we able to use this in different pedagogical settings?

Whether or not the written word or an image can speak is, of course, a matter of how, not only silence, but also the concept of speech are defined. Perhaps it is better to use the term *expression* instead—various symbolic systems express something that is conveyed to the observer. Something that is quite clear, however, is that Dewey (1991), Merleau-Ponty (1995) and others believe that there is something beyond the spoken word that can appear in connection with visualisation. An image conveys a message.

At school, the spoken word is often highly valued, and the written word perhaps even more so. Students must, however, speak and write at the right place, at the right time and about the right things. It is worth taking a moment here to consider whether a picture is valued equally with the spoken or written word. We can also think about whether these nonverbal modes of expression are silent or not. If they are silent, does the teacher in that case have the opportunity to hear them?

Italian philosopher and educator Reggio Emilia builds on the basic assumption that a child has a hundred languages it can use to express and communicate its observations and experiences (Edwards, Gandini, & Forman, 1998). Most of these languages, though, are by nature nonverbal and to some extent silent, but are nevertheless able to express experiences and feelings. Although we humans—according to Reggio Emilia—have a hundred languages, we usually only use one or two of them. Speaking is the most common way most people communicate with each other, and it appears to be the most common form of communication in educational settings such as schools. By using different modes of expression, however, conditions for learning are improved. In schools, it is therefore important to be observant of the fact that students vary as to their ability to express their observations and experiences. Rather than expressing themselves verbally, some would perhaps prefer to use writing or other symbolic language—so-called nonverbal modes of expression. Students should therefore be given the opportunity to use as many different languages as possible to express themselves.

Silence can be viewed as a condition where sound is absent, and in that light, several of the nonverbal modes of expression are silent. Although silent, they still convey something well worth considering, listening to. So, from that perspective, they are not silent at all. Regardless of how these different symbolic systems are viewed—as silent or otherwise—they can be used as tools for transmitting observations and experiences, which can, in turn, show the importance of using nonverbal modes of expression in educational settings as a complement to verbal language.

ACTIVITY VERSUS STILLNESS IN THE SCHOOL WORLD

If we bear in mind that people are capable of silent language and possess silent knowledge, questions arise about whether and how silence is valued and treated in places such as schools. In other words, what significance does silence have in different pedagogical settings? I have described the paradoxical situation that arises in various educational settings where, on the one hand, students talk and make noise when, according to the teacher, they should be quiet, but on the other hand, they do not comment or share opinions in discussions and oral activities when everyone is expected to express themselves verbally. Is a student's silence accepted, or does the school see it as a problem?

Silence is important in educational settings, which is especially apparent in the phrase 'Silence in class.' But the meaning behind those words can change according to how they are emphasised. Something else that changes the meaning of the words, which also indicates emphasis, is when the phrase is followed by an exclamation mark, a question mark or perhaps a full stop. Permit me to illustrate such a change in the meaning of the words: 'Silence in class!' the teacher orders the students. Compare that with how it became 'silent in class' when the students found out about a classmate's illness. The same words convey a completely different message, which in turn affects the meaning of the silence itself. Silence manifests itself in different forms.

Another aspect of silence in connection with school activities that we should think about is the importance of a response or feedback from teacher to student, and sometimes the other way around. A student who submits a written assignment, or perhaps a written test, expects to receive some sort of response. If it is not received, it could be interpreted in different ways. Of course, this applies not only to written assignments, but also to oral presentations and so on. A presentation that is greeted with silence can be perceived by those who gave it as unpleasant and uncomfortable: 'Did I (we) say something wrong?,' 'Have I (we) made a fool of myself (ourselves)?' But perhaps the silence occurred because what was just said gave rise to contemplation and reflection, and those who listened need time and space in stillness to allow the message to sink in. If that is the case, then the silence might instead be perceived as both pleasing and comfortable.

Even if no words are spoken in both of these situations, small silent nuances of body language amongst the audience may determine how the silence is perceived by those who have just spoken: a quiet look that conveys either contempt or admiration, an indifferent shrug, or a quiet, affirmative nodding, a person's position on their chair, or perhaps the position of the arms.

Stopping to think about silence can draw attention to the importance of being allowed the opportunity to be in, and also to listen to, silence in different pedagogical settings. The use of silence, such as dramatic pauses, can create better conditions for learning than can oral repetitions, however powerfully they are delivered. Teaching can thus occur in silence, with silence and through silence. We should also learn about silence so as to be prepared to create good conditions for teaching and learning.

WHAT IS SAID IN THAT WHICH IS NOT SAID?

With the support of Bateson, Dickinson, Dewey and others, I have argued that silence can convey a message. This was highlighted through the discussion on silent communication in questionnaires. One aspect worth considering in connection with silent or silenced messages in questionnaires is that, even though students express their thoughts, experiences, feelings or opinions by responding to a survey, their voices are not always listened to. I would like to highlight the question of whether students who complete a questionnaire feel that someone is really listening to them and taking their responses seriously. For example, we should think about how students receive feedback from those who distributed, collected, processed and reported the results of a survey. Perhaps it is here that silence speaks its clear language.

A further aspect worth highlighting in connection with silent or silenced messages in a survey is the trend of having people fill in a survey online, such as through Computer-Assisted Personal Interviewing and Computer-Assisted Self Interviewing. Bradburn, Frankel and Baker (1991) argue that, when a questionnaire is filled in online, the respondent's view of their task changes. The medium changes their understanding of what the task is, which can, in turn, affect how the person in question acts. Amongst other things, these researchers reason that the response rate increases and that extra information does not get written in the margins when online surveys are used—it is simply not possible to append additional information in some of the questionnaires that are distributed and responded to online. In some cases, the system also alerts the user if a question has not been answered. Refraining from selecting an answer to a survey question is therefore not an option. On this subject, it is appropriate to think about how accurate and truthful a response becomes when the person answering a question does not really want, or is not able, to answer but is forced by the system to select an answer.

The question is whether this is relevant in the context of a discussion on silence. Silent communication in a survey can occur regardless of whether the questionnaire is completed in paper or online. Instead, I wish to highlight the importance of being alert and taking silent or silenced messages that are often included in surveys into consideration, all in an effort to gain as complete a picture as possible of what is requested. And perhaps, by listening to the silence, we will acquire even more valuable information than what is requested, or possibly even something completely different of value in connection with what is being asked.

Earlier, I argued that silence, or that which cannot be spoken, can be just as important as what is able to be said. Based on this argument, a mark beside or between checkboxes in a questionnaire, unanswered questions or statements, and/or written comments in the margins could be considered to be every bit as important as the answers provided within the preprinted boxes. It could even be more important, and perhaps, as the expression goes, silence is golden.

Talking About Silence

In the foreword to this book, I asked whether it is possible to talk about silence. Now that we are approaching the end of the text, it is time to again consider this question. Heidegger even believed that talking and writing about silence produces intolerable babble, and there is, perhaps, some truth in that. Perhaps silence as a phenomenon needs and presupposes exactly that—silence. No silence without silence. Whether or not this book is merely intolerable babble is something for others to evaluate, but according to my horizon of understanding, it is entirely possible to talk, and even write, about certain aspects of silence, though I do not claim that these are the only aspects that apply. Perhaps certain other aspects of silence need to be experienced without words—we have to be in silence. Although we can talk about silence in silence, as soon as we begin to speak, we are no longer in silence; the silence is broken by the spoken word.

However, I wish to argue that one way of contributing to awareness of the many facets and forms of silence is to talk about them. Another way is to write about them. Putting words to silence, regardless of whether those words are spoken aloud or written down, can help the listener or reader to begin to think and reflect on silence.

We humans are able to convey messages without speaking—silence says something. Dickinson (2020) goes so far as to say that silence says more than the spoken word. In my opinion, the important thing is not to try and work out which message says the most, nor is it to decide which one is best. Instead, what is important is to be aware of, and benefit from, the various aspects and forms of silence. This becomes especially important in the many educational situations that occur every day in our schools.

Silence exists as a natural and undeniable part of our daily lives—in some sense, we can never completely avoid silence. On the other hand, the way silence is perceived can vary greatly. Different people can perceive silence in completely different ways, as has been shown in this publication. Is silence longed for and appreciated, or is it perhaps forced and intimidating? Silence can, however, be much more nuanced than that, and Bollnow (1982) emphasises that silence between two people can sometimes indicate an understanding between them, without any words being spoken. This is a kind of wordless agreement, where people understand each other without needing to use the spoken word. One way of describing this silent understanding between two people is as follows: 'It is the mutual silence that has the sensation of belonging together—a good friend is the one you can sit with in silence' (Alerby & Elídóttir, 2003, p. 42). Perhaps most important of all is to not be afraid of silence, but to instead dare to listen to it and use it in a constructive way.

FINALLY

Sometimes, it is better to leave things unsaid, or, in this case, unwritten, instead of specifically spelling everything out; to dare to leave the text silent. It is my hope that even the silent spaces between these words should convey something to you, dear reader. Because, just as van Manen points out, a written text can to some extent be both unspoken—silent—and explanatory. But that does require the reader to be responsive, sensitive and open. Or, to put it another way, receptive to the silence surrounding the words. It follows that it has not been my intention to comprehensively ascertain what silence is in its full potential. Instead, my aim has been that this text should encourage contemplation and reflection on silence—that the text you have just read should be completed by you, the reader, as you fall into silence through the stillness of reflection. To this end, I have collected together a few different aspects, meanings, examples and forms of silence, which I have highlighted and discussed. This, in turn, means that certain other aspects, meanings, examples and forms have been left in silence.

Is it possible to accumulate silence? In *Murke's Collected Silences*, a short story by Heinrich Böll (1995), the protagonist, Dr Murke, collects the silent moments and pauses that have been cut out of various radio recordings that he is assigned to edit at the radio station where he works. When he comes home from a long day of soul-destroying work, he listens to the recorded silence he has on his tapes. In an attempt to escape the noise of having to listen to unbearable radio lectures and remove certain words or edit some text from four hours down to four minutes, the silence of the tape recordings becomes a kind of refuge to him. We can, perhaps, draw parallels between Dr Murke's life and today's school students and ask the questions: How much noise (and perhaps even soul-destroying assignments) are students exposed to during a school day? Is there a silent refuge available to students during their day at school, and, if so, what is it like? Perhaps such a quiet refuge can be found in the example described earlier from Australia—the Peace Area. There, students escape from the noise to find peace and stillness. In this context, we might also spare a thought for all the teachers who work out there in our schools. How is their noise level? Do teachers have access to a quiet and tranquil place during the school day? And perhaps they, too, like Dr Murke, have to endure the odd soul-destroying assignment. Whatever the case may be, both students and teachers have to deal with the school itself as well as the activities that are arranged there, which all occur within designated time frames. To have the opportunity to withdraw within this predetermined framework, to be able to sit by oneself in silence without having to leave the school building or grounds, is of utmost importance to all the children, adolescents and adults located and working there. If school is to be the best of the best when it comes to creating conditions for knowledge and learning, formation and transformation and engagement and curiosity (the good), silence is needed, along with opportunity to retreat in solitude (cf. Stern, 2016).

Silence is an essential and unavoidable part of education and life in general. It is therefore important to gain an awareness and understanding of silence and its significance. We are unlikely to ever fully understand silence, even though the Swedish writer, historian, poet, philosopher and composer Erik Gustav Geijer wrote that: 'The only wise one, is the one who understands silence' (Geijer, 1999, p. 204). But if we fall short in understanding silence, does that mean we are lacking in wisdom? I choose instead to conclude by paraphrasing that verse in the poem by Erik Gustav Geijer: 'The only wise one, is the one who *appreciates* silence.'

REFERENCES

Adelmann, K. (2002). *Att lyssna till röster: ett vidgat lyssnandebegrepp i ett didaktiskt perspektiv* [Listening to voices. An extended notion of listening in an educational perspective] Doctoral thesis, Malmö högskola, Malmö.

Adelmann, K. (2008). *Konsten att lyssna. Didaktiskt lyssnande i skola och utbildning. The art of listening* [The art of listening. Didactic listening in school and education] Lund, Sweden: Studentlitteratur.

Alerby, E., & Elídóttir, J. (2003). The sounds of silence: Some remarks on the value of silence in the process of reflection in relation to teaching and learning. *Reflective Practice, 4*(1), 41–51.

Babisch, W., Fromme, H., Beyer, A., & Ising, H. (2001). Increased catecholamine levels in urine in subjects exposed to road traffic noise: The role of stress hormones in noise research. *Environment International, 26*(7–8), 475–481.

Bateson, G. (1987). *Steps to ecology of mind*. Northvale, NJ: Jason Aronson Inc.

Böll, H. (1995). *Murke's collected silences*. Evanston, IL: Northwestern University Press.

Bollnow, O. F. (1982). On silence – Findings of philosophicopedagogical anthropology. *Universitas, 24*(1), 41–47.

Boyd, D. (2007). Why youth (heart) social network sites: The role of networked publics in teenage social life. In D. Buckingham (red.), *Mac Arthur foundation series on digital learning – Youth identity, and digital media volume*. Cambridge, MA: MIT Press.

Boyd, D., & Ellison, N. (2008). Social network sites: Definition, history and scholarship. *Journal of Computer-Mediated Communication, 13*, 210–230.

Bradburn, N., Frankel, M., & Baker, R. (1991, maj 16–19). *A comparison of computer-assisted personal interviews (CAPI) with paper-and-pencil (PAPI) interviews in the national longitudinal study of youth*. Paper presenterat på AAPOR-konferensen, Phoenix, Arizona.

Cooper, D. E. (2012). Silence, nature and education. In H. Fiskå Hägg & A. Kristiansen (Eds.), *Attending to silence. Educators and philosophers on the art of listening*. Kristiansand, Norway: Portal Academic.

Dewey, J. (1991). *How we think*. New York: Prometheus Books.

Dickinson, E. (2020). Emily Dickinson. https://www.poetryfoundation.org/poets/emily-dickinson (accessed 2 July 2020).

Edwards, C., Gandini, L., & Forman, G. (1998). *The hundred languages of children: The Reggio Emilia approach – Advanced reflections*. Greenwich, UK: Ablex Publication.

Englund, A. (2000). *Trafikstress. En redovisning av utförande och resultat i KFB-finansierade forskningsprojekt 1996–1999* [Traffic Stress. A description of the performance and results in the KFB funded research 1996–1999]. Stockholm: Fritzes.

Englund, P. (2005). *Tystnadens historia och andra essäer* [The history of silence and other essays]. Stockholm: Publishing House Atlantis AB.

Geijer, E.G. (1999). *Dikter* [Poems]. Stockholm: Atlantis.

Hedquist, R. (2006). Lyssna och lära [Listen and learn]. *Aktum, 6*, 24.

Maitland, S. (2009, orig. 2008). *A book of silence. A journey in search of the pleasures and powers of silence.* London: Granata Books.

Merleau-Ponty, M. (1995). *Signs.* Evanston, IL: Northwestern University Press.

Stern, J. (2016). Solitude and spirituality in schooling: The alternative at the heart of the school. In H.E. Lees, & N. Noddings (Eds.), *The Palgrave international handbook of alternative education* (pp. 431–445). London: Palgrave Macmillan. Chapter 28.

Swedish Work Environment Authority (2013). Bullrig arbetsmiljö sänker prestationsnivån i skolan [Noisy work environment lowers the level of performance in school]. https://www.av.se/. Accessed 23 Apr 2020.

Von Wright, M. (2012). Silence in the asymmetry of educational relations. In H. Fiskå Hägg & A. Kristiansen (Eds.), *Attending to silence. Educators and philosophers on the art of listening.* Kristiansand, Norway: Portal Academic.

PREVIOUS PUBLICATIONS AND PRESENTATIONS ON SILENCE

Alerby, E. (2004). The appreciation of a quiet place at school. *Didaktisk tidskrift, 14*(1), 57–61.

Alerby, E. (2002a, May 7–9). *Silence as an essential notion within the educational field.* Paper presented at NERA:s Congress, Tallinn, Estland.

Alerby, E. (2002b). Några betraktelser av tystnadens betydelse i pedagogiska sammanhang. *VÅRD, 3,* 32–37.

Alerby, E. (2007, March 15–17). *The notion of silence in educational context.* Paper presented at NERA:s Congress, Wasa, Finland.

Alerby, E. (2010, June 14–15). *Silence in education.* Paper presented at the Third Seminar of Silence: Silent Revisited, Esbo, Finland.

Alerby, E. (2012a). *Om tystnad – i pedagogiska sammanhang* [About silence – In pedagogical settings]. Lund, Sweden: Studentlitteratur.

Alerby, E. (2012b). Silence within and beyond education. In H. Fiskå Hägg & A. Kristiansen (Eds.), *Attending to silence. Educators and philosophers on the art of listening.* Kristiansand, Norway: Portal Academic.

Alerby, E. (2012c, May 2–4). *Rethinking the schoolyard as a place for silence.* Full reviewed paper presented at the conference Philosophical Perspectives in Outdoor Education, University of Edinburgh, Scotland.

Alerby, E. (2012d, March 8–10). *About silence – A matter for educational settings.* Paper presented at NERA:s Congress, Copenhagen, Denmark.

Alerby, E. (2019a). Places for silence and stillness in schools of today: A matter for educational policy. *Policy Futures in Education, 17*(4), 530–540.

© The Author(s) 2020   89
E. Alerby, *Silence within and beyond Pedagogical Settings,*
https://doi.org/10.1007/978-3-030-51060-2

Alerby, E. (2019b, April 10–12). *Silence, senses and solitude in the light of texture.* Paper presented at the International Pandisciplinary Symposium on Solitude in Community – Alone Together, York St John University, York, UK.

Alerby, E. (2019c, March 5–6). *Texture: In the light of senses, silence and the lived body.* Invited paper at the pre-conference 'Educating the Senses in a Globalized World', The Nordic Society for Philosophy of Education, Uppsala University, Sweden.

Alerby, E. & Elídóttir, J. (2003). The sounds of silence: Some remarks on the value of silence in the process of reflection in relation to teaching and learning. *Reflective Practice, 4*(1), 41–51.

Alerby, E., & Hertting, K. (2012, December 2–6). *'Here I am, here I can think': The significance of the place and silent dimensions of knowledge.* Paper presented at the AARE Conference (Australian Association for Research in Education), Sydney.

Alerby, E., & Kostenius, C. (2011a, July 27–30). *Silence for health and learning – A phenomenological reflection.* Paper presented at the 30:e International Human Science Research Conference, Oxford, UK.

Alerby, E., & Kostenius, C. (2011b). 'Dammed Taxi Cab' – How silent communication in questionnaires can be understood and used to give voice to children's experiences. *International Journal of Research & Method in Education, 34*(2), 1–14.

Alerby, E., & Kostenius-Foster, C. (2005, March 10–12). *A silent message is also a message.* Paper presented at NERA:s Congress, Oslo, Norge.

Alerby, E., & Westman, S. (2013, December 6–9). *A critical exploration of assessment in relation to silence and silent students.* Full referee paper presented at the PESA conference (Philosophy of Education Society of Australasia), Melbourne University, Australia.

Doseth Opstad, K., & Alerby, E. (2017). *Textur, tystnad och kroppslighet* [Texture, silence and embodiment]. Paper presented at the e17 conference: 'Tacit knowing or just plain silence?', 31 October–2 November 2017, Umeå, Sweden.

REFERENCES

Aasland, D. G. (2012). Coming home to silence. In H. Fiskå Hägg & A. Kristiansen (Eds.), *Attending to silence. Educators and philosophers on the art of listening*. Kristiansand, Norway: Portal Academic.

Achino-Loeb, M.-L. (red.). (2006). *Silence. The currency of power.* New York: Berghahn Books.

Adelmann, K. (2002). *Att lyssna till röster: ett vidgat lyssnandebegrepp i ett didaktiskt perspektiv* [Listening to voices. An extended notion of listening in an educational perspective] Doctoral thesis, Malmö högskola, Malmö.

Adelmann, K. (2008). *Konsten att lyssna. Didaktiskt lyssnande i skola och utbildning. The art of listening* [The art of listening. Didactic listening in school and education] Lund, Sweden: Studentlitteratur.

Ahonen, A. (2010). *Psychosocial well-being of schoolchildren in the Barents region. A comparison from the northern parts of Norway, Sweden and Finland and Northwest Russia*. Doctoral thesis, University of Lapland Press, Rovaniemi.

Ahonen, A., Alerby, E., Johansen, O.-M., Rajala, R., Ryzhkova, I., Sohlman, E., & Villanen, H. (red.). (2008). *Crystals of schoolchildren's well-being. Cross-border training material for promoting psychosocial well-being through school education*. Rovaniemi, Finland: University of Lapland Press.

Alerby, E. (2002a, May 7–9). *Silence as an essential notion within the educational field*. Paper presented at NERA:s Congress, Tallinn, Estland.

Alerby, E. (2002b). Några betraktelser av tystnadens betydelse i pedagogiska sammanhang. *VÅRD, 3*, 32–37.

Alerby, E. (2003). 'During the break we have fun.' A study concerning pupils' experience of school. *Educational Research, 45*(1), 17–28.

Alerby, E. (2004). The appreciation of a quiet place at school. *Didaktisk tidskrift, 14*(1), 57–61.

© The Author(s) 2020   91
E. Alerby, *Silence within and beyond Pedagogical Settings*,
https://doi.org/10.1007/978-3-030-51060-2

Alerby, E. (2007, March 15–17). *The notion of silence in educational context*. Paper presented at NERA:s Congress, Wasa, Finland.

Alerby, E. (2010, June 14–15). *Silence in education*. Paper presented at the Third Seminar of Silence: Silent Revisited, Esbo, Finland.

Alerby, E. (2012a, March 8–10). *About silence – A matter for educational settings*. Paper presented at NERA:s Congress, Copenhagen, Denmark.

Alerby, E. (2012b, May 2–4). *Rethinking the schoolyard as a place for silence*. Full reviewed paper presented at the conference Philosophical Perspectives in Outdoor Education, University of Edinburgh, Scotland.

Alerby, E. (2012c). *Om tystnad – i pedagogiska sammanhang* [About silence – In pedagogical settings]. Lund, Sweden: Studentlitteratur.

Alerby, E. (2012d). Silence within and beyond education. In H. Fiskå Hägg & A. Kristiansen (Eds.), *Attending to silence. Educators and philosophers on the art of listening*. Kristiansand, Norway: Portal Academic.

Alerby, E. (2015). 'A picture tells more than thousand words.' Drawings used as research method. In J. Brown & N. Johnson (Eds.), *Children's images of identity. Drawing the self and the other*. Rotterdam, Netherlands: Sense Publisher.

Alerby, E. (2019a). Places for silence and stillness in schools of today: A matter for educational policy. *Policy Futures in Education, 17*(4), 530–540.

Alerby, E. (2019b, March 5–6). *Texture: In the light of senses, silence and the lived body*. Invited paper at the pre-conference "Educating the Senses in a Globalized World", The Nordic Society for Philosophy of Education, Uppsala University, Sweden.

Alerby, E. (2019c, April 10–12). *Silence, senses and solitude in the light of texture*. Paper presented at the International Pandisciplinary Symposium on Solitude in Community – Alone Together, York St John University, York, UK.

Alerby, E., & Elídóttir, J. (2003). The sounds of silence: Some remarks on the value of silence in the process of reflection in relation to teaching and learning. *Reflective Practice, 4*(1), 41–51.

Alerby, E., & Hertting, K. (2012, December 2–6). *'Here I am, here I can think': The significance of the place and silent dimensions of knowledge*. Paper presented at the AARE Conference (Australian Association for Research in Education), Sydney.

Alerby, E., & Kostenius, C. (2011a, July 27–30). *Silence for health and learning – A phenomenological reflection* . Paper presented at the 30:e International Human Science Research Conference, Oxford, UK.

Alerby, E., & Kostenius, C. (2011b). 'Dammed Taxi Cab' – How silent communication in questionnaires can be understood and used to give voice to children's experiences. *International Journal of Research & Method in Education, 34*(2), 1–14.

Alerby, E., & Kostenius-Foster, C. (2005, March 10–12). *A silent message is also a message*. Paper presented at NERA:s Congress, Oslo, Norge.

Alerby, E., & Westman, S. (2013, December 6–9). *A critical exploration of assessment in relation to silence and silent students*. Full referee paper presented at the PESA conference (Philosophy of Education Society of Australasia), Melbourne University, Australia.

Arbetsmiljöverket. (2010). www.av.se. 25 July 2010.

Archer, A. (2001). *A beginner's guide to Japan*. www.shinnova.com/part/99-japa/abj17-e.htm. 7 Dec 2001.

Arendt, H. (1958). *The human condition*. Chicago: Chicago University Press.

Arlinger, S. (1995). Det utsatta örat [The vulnerable ear]. In H. Karlsson (Ed.), *Svenska ljudlandskap. Om hörseln, bullret och tystnaden* [Swedish soundscape. About the hearing, the noise and the silence]. Göteborg, Sweden: Bo Ejeby Förlag.

Arnheim, R. (1969). *Visual thinking*. Berkeley, CA: University of California Press.

Babisch, W., Fromme, H., Beyer, A., & Ising, H. (2001). Increased catecholamine levels in urine in subjects exposed to road traffic noise: The role of stress hormones in noise research. *Environment International, 26*(7–8), 475–481.

Bann, S. (1967). *Concrete poetry: An international anthology*. London: London Magazine.

Barnard, W. F. (1913). *The tongues of toil and other poems*. Chicago: The Workers' Art Press.

Bateson, G. (1987). *Steps to ecology of mind*. Northvale, NJ: Jason Aronson Inc.

Bell, R., & Battin, B. (1995). *Seeds of the Spirit: Wisdom of the twentieth century*. Louisville, OH: Westminster John Knox Press.

Bengtsson, J. (1993). *Sammanflätningar. Husserls och Merleau-Pontys fenomenologi* [Intertwinings. The phenomenology of Husserl and Merleau-Ponty]. Göteborg, Sweden: Daidalos AB.

Bergmark, U., & Kostenius, C. (2009). 'Listen to me when I have something to say' – Students' participation in research for sustainable school improvement. *Improving Schools, 12*(3), 249–260.

Bergmark, U., & Kostenius, C. (2012). Student visual narratives giving voice to positive learning experiences – A contribution to educational reform. *Academic Leadership Journal, 10*(1), 1–17.

Böll, H. (1995). *Murke's collected silences*. Evanston, IL: Northwestern University Press.

Bollnow, O. F. (1982). On silence – Findings of philosophicopedagogical anthropology. *Universitas, 24*(1), 41–47.

Boucher, T. (2002). *Pantomimteatern*. www.pantomimteatern.com/om-oss/historia/mimkonsten/. 19 Jan 2011.

Bourdieu, P. (1993). *Sociology in question*. London: Sage.

Boverket. (2011). *Vad är ljud och buller?* [What is sound and noise?] www.boverket.se/Planera/planeringsfragor/Buller/Vad-ar-ljud-och-buller/. 9 Feb 2011.

Boyd, D. (2007). Why youth (heart) social network sites: The role of networked publics in teenage social life. In D. Buckingham (red.), *Mac Arthur foundation series on digital learning – Youth identity, and digital media volume.* Cambridge, MA: MIT Press.

Boyd, D., & Ellison, N. (2008). Social network sites: Definition, history and scholarship. *Journal of Computer-Mediated Communication, 13,* 210–230.

Bradburn, N., Frankel, M., & Baker, R. (1991, maj 16–19). *A comparison of computer-assisted personal interviews (CAPI) with paper-and-pencil (PAPI) interviews in the national longitudinal study of youth.* Paper presenterat på AAPOR-konferensen, Phoenix, Arizona.

Brown, L. (red.). (1993). *The new shorter Oxford English dictionary. On historical principles.* Oxford: Oxford University Press.

Bruner, J. (1996). *The culture of education.* Cambridge, MA: Harvard University Press.

Buber, M. (1978). *Between man and man.* New York: Macmillan Publishing Co.

Buber, M. (1988). *The knowledge of the man: Selected essays.* Atlantic Highlands, NJ: Humanities Press International, Inc.

Buber, M. (1993). *Dialogens väsen: traktat om det dialogiska livet* [The essence of dialogue: Treaty on dialogue life]. Ludvika, Sweden: Dualis.

Cage, J. (1961). *Silence: Lectures and writings.* Middletown, CT: Wesleyan University Press.

Cage, J. (1997). *I-VI John Cage.* Hanover, NH: University Press of New England.

Cohen, L. (1992). 'Anthem' in the music album *The Future.* https://www.leonardcohen.com/. Accessed 16 Jan 2020.

Collins English Dictionary. (1992). Glasgow, Scotland: Harper Collins Publishers.

Cooper, D. E. (2008). *The philosophy of gardens.* Oxford: Oxford University Press.

Cooper, D. E. (2012). Silence, nature and education. In H. Fiskå Hägg & A. Kristiansen (Eds.), *Attending to silence. Educators and philosophers on the art of listening.* Kristiansand, Norway: Portal Academic.

Dewey, J. (1991). *How we think.* New York: Prometheus Books.

Dickinson, E. (2020). Emily Dickinson. https://www.poetryfoundation.org/poets/emily-dickinson. Accessed 2 July 2020.

Doseth Opstad, K., & Alerby, E. (2017). *Textur, tystnad och kroppslighet* [Texture, silence and embodiment]. Paper presented at the e17 conference: 'Tacit knowing or just plain silence?', 31 October–2 November 2017, Umeå, Sweden.

Dysthe, O. (1993). *Writing and talking to learn. A theory-based, interpretive study in three classrooms in the USA and Norway.* Tromsø, Norway: University of Tromsø.

Edwards, C., Gandini, L., & Forman, G. (1998). *The hundred languages of children: The Reggio Emilia approach – Advanced reflections.* Greenwich, UK: Ablex Publication.

Eggeby, E., & Söderberg, J. (1999). *Kvantitativa metoder: för samhällsvetare och humanister* [Quantitative methods: For social sciences and humanities]. Lund, Sweden: Studentlitteratur.

Eisner, E. W. (1997). The promise and perils of alternative forms of data representation. *Educational Researcher, 26*(6), 4–10.

Englund, A. (2000). *Trafikstress. En redovisning av utförande och resultat i KFB-finansierade forskningsprojekt 1996–1999* [Traffic Stress. A description of the performance and results in the KFB funded research 1996–1999]. Stockholm: Fritzes.

Englund, P. (2005). *Tystnadens historia och andra essäer* [The history of silence and other essays]. Stockholm: Publishing House Atlantis AB.

EUdict dictionary. (2019). Cum tacent clamant. http://eudict.com. 3 Nov 2019.

Forsman, A. (2003). *Skolans texter mot mobbning – reella styrdokument eller hyllvärmare?* [School texts against bullying – Real policy documents or shelving heaters?]. Doctoral thesis, Luleå University of Technology, Luleå.

Foucault, M. (1978). *The history of sexuality: Volume 1: An Introduction.* New York: Pantheon Books.

Freire, P. (1972). *Pedagogy of the oppressed.* London: Penguin Books.

Gadamer, H.-G. (1975). *Truth and method.* New York: Seabury.

Gann, K. (2010). *No such thing as silence. John Cage's 4'33".* London: Yale University Press.

Geijer, E.G. (1999). *Dikter* [Poems]. Stockholm: Atlantis.

Gunnarsdottir, H., Bjereld, Y., Hensing, G. & Petzold, M. (2015). Associations between parents' subjective time pressure and mental health problems among children in the Nordic countries: a population based study. *BMC Public Health, 15*(353). https://doi.org/10.1186/s12889-015-1634-4.

Gutmann, P. (1999). *John Cage and the Avant-Garde: The sounds of silence.* www.classicalnotes.net/columns/silence.html. 20 Jan 2011.

Halberstadt event. (2011). www.john-cage.halberstadt.de. 22 Jan 2011.

Harvey, P. (2000). *An introduction to Buddhist ethics: Foundations, values and issues.* Cambridge, MA: Cambridge University Press.

Hedquist, R. (2006). Lyssna och lära [Listen and learn]. *Aktum, 6,* 24.

Heidegger, M. (1971). *On the way to language.* San Francisco: Harper Collins Publisher.

Heidegger, M. (1992). *History of the concept of time.* Bloomington, IN: Indiana University Press.

Heidegger, M. (1993). *Varat och tiden* [Being and time]. Göteborg, Sweden: Daidalos.

Hellberg, A. (red.). (2002). *Buller och bullerbekämpning* [Noise and noise control]. Stockholm: Arbetsmiljöverket.

Holm, P. (1976). *Bevingade ord* [Winged words]. Stockholm: Albert Bonniers förlag.

Husserl, E. (1995). *Fenomenologins idé* [The idea of phenomenology]. Göteborg, Sweden: Daidalos.

Jaworski, A. (1993). *The power of silence. Social and pragmatic perspectives.* London: Sage.

Jensen, V. (1973). Communicative functions of silence. *ETC, 30,* 249–257.

Kellett, M. (2010). Small shoes, big steps! Empowering children as active researchers. *American Journal of Community Research, 46*, 195–203.

King, A., Wold, B., Tudor-Smith, C., & Harel, Y. (1996). *The health of youth. A WHO cross national survey* (European series 69). WHO Regional Publications.

Kostelanetz, R. (1988). *Conversing with Cage*. New York: Limelight.

Kostenius, C. (2008). *Giving voice and space to children in health promotion*. Doctoral thesis, Luleå University of Technology, Luleå.

Kress, G., & van Leeuwen, T. (2001). *Multimodal discourse. The modes and media of contemporary communication*. London: Arnold Publishers.

Lebra, T. S. (1987). The cultural significance of silence in Japanese communication. *Multilingua, 6*(4), 343–357.

Lees, H. (2012). *Silence in schools*. London: Trentham Books.

Loan, O. (2002). *The elements of design: Rediscovering colors, textures, forms and shapes*. London: Thames & Hudson.

Lögstrup, K. E. (1975). *Den etiske fordring* [The ethical requirement]. Köpenhamn, Denmark: Gyldendal.

Loughran, J. (1996). *Developing reflective practice: Learning about teaching and learning through modelling*. London: Falmer Press.

Maitland, S. (2009, orig. 2008). *A book of silence. A journey in search of the pleasures and powers of silence*. London: Granata Books.

Mendes-Flohr, P. (2012). Dialogical silence. In H. Fiskå Hägg & A. Kristiansen (Eds.), *Attending to silence. Educators and philosophers on the art of listening*. Kristiansand, Norway: Portal Academic.

Merleau-Ponty, M. (1995). *Signs*. Evanston, IL: Northwestern University Press.

Merleau-Ponty, M. (1996). *Phenomenology of perceptions*. London: Routledge.

Ministry of Education and Research. (1994). *Curriculum for the compulsory school system, pre-school class and the leisure-time centre—Lpo94*. Stockholm: Fritzes.

Ministry of Education and Research. (2011). *Curriculum for the compulsory school, preschool class and school-age educare—Lgr11*, Revised 2018. Stockholm: Fritzes.

Natur och Kulturs Svenska Ordbok. (2001). Stockholm: Publishing House Natur och Kultur.

Norstedts etymologiska ordbok. (2008). Stockholm: Norstedts Akademiska förlag.

Nygren, L., & Blom, B. (1999). Analys av korta narratives [Analysis of short narratives]. In J. Lidén, G. Westlander, & G. Karlsson (red.), *Kvalitativa metoder i arbetslivsforskning* [Qualitative methods in working life research]. Uppsala, Sweden: Rådet för arbetslivsforskning.

Opstad, K. D. (1990). *Teksturer i vev* [Textures in tissues]. Master thesis, Statens lærerhøgskole i forming Oslo.

Opstad, K.D (2010). Estetisk dannelse – estetiske fags bidrag i skolens dannelsesperspektiv [Aesthetic education – Aesthetic subjects' contribution to the formation perspective of school]. In M. Brekke (Ed.), *Dannelse i skole og utdanning* [Formation in school and education]. Oslo, Norway: Universitetsforlaget.

Passchier-Vermeer, W., & Passchier, W. F. (2000). Noise exposure and public health. *Environmental Health Perspectives, 108*(1), 123–131.

Pittenger, R. E., Hockett, C. F., & Danehy, J. J. (1960). *The first five minutes: A sample of microscopic interview analysis.* Ithaca, NY: Paul Martineau.

Polanyi, M. (1958). *Personal knowledge.* Chicago: University of Chicago Press.

Polanyi, M. (1969). *Knowing and being.* Chicago: University of Chicago Press.

Rönnbäck, J. (2004). *Politikens genusgränser: den kvinnliga rösträttsrörelsen och kampen för kvinnors politiska medborgarskap 1902–1921* [Political gender boundaries: The women's suffrage movement and the struggle for women's political citizenship 1902–1921]. Doctoral thesis, Atlas, Stockholm.

Rowe, M. B. (1974). Pausing phenomena: Influence on the quality of instruction. *Journal of Psycholinguistic Research, 3*(3), 203–224.

Rowe, M. B. (1986). Wait time: Slowing down may be a way of speeding up. *Journal of Teacher Education, 37*, 43–50.

Schultz, K. (2010). After the Blackbird Whistles: Listening to Silence in Classrooms. Teachers College Record. The Voice of Scholarship in Education.

Schütz, A. (1999). *Den sociala världens fenomenologi* [The phenomenology of the social world]. Göteborg, Sweden: Daidalos.

Schwarz, N., & Strack, F. (1991). Context effects in attitude surveys: Applying cognitive theory to social research. In W. Stroebe & M. Hewstone (red.), *European Review of Social Psychology* (Vol. 2, pp. 31–50).

Schwarz, N., & Strack, F. (1999). Reports of subjective well-being: Judgmental processes and their methodological implications. In D. Kahneman, E. Diener, & N. Schwarz (red.), *Well-being: The foundations of hedonic psychology.* New York: Russell Sage Foundation.

Scollon, R., & Scollon, S. (1987). *Responsive communication.* Haines, AK: Black Current Press.

Simpson, J.A., & Weiner, E.S.C (red.). (1989). *The Oxford English Dictionary, second edition, Volume XV, Ser–Soosy.* Oxford: Clarendon Press.

Skantze, A. (1989). *Vad betyder skolhuset? Skolans fysiska miljö ur elevernas perspektiv studerad i relation till barns och ungdomars utvecklingsuppgifter* [What is the significance of the school building: The physical environment of the school seen from a pupil perspective and studied in relation to their developmental tasks]. Doctoral thesis, Stockholm University, Stockholm.

Smyth, J. (2006a). 'When students have power': Student engagement, student voice, and the possibilities for school reform around 'dropping out' of school. *International Journal of Education in Leadership, 9*(4), 279–284.

Smyth, J. (2006b). Educational leadership that foster students' voice. *International Journal of Education in Leadership, 9*(4), 285–298.

Solt, M. E. (1968). *Concrete poetry: A world view.* Bloomington, IN: Indiana University Press.

Sörlin, S. (2004). *Världens ordning* [The order of the world]. Stockholm: Publishing House Natur och Kultur.

Southwold, M. (1983). *Buddhism in life: The anthropological study of religion and Sinhalese practice of Buddhism*. Manchester, England: Manchester University Press.

Stern, J. (2014). *Loneliness and solitude in education: How to value individuality and create an Enstatic school*. Oxford: Peter Lang.

Stern, J. (2016). Solitude and spirituality in schooling: The alternative at the heart of the school. In H.E. Lees, & N. Noddings (Eds.), *The Palgrave international handbook of alternative education* (pp. 431-445). London: Palgrave Macmillan. Chapter 28.

Swedish National Agency for Education. (2010a). *Kursplan för moderna språk* [Syllabus for modern languages]. www.skolverket.se/sb/d/2386/a/16138. 22 July 2010.

Swedish National Agency for Education. (2010b). *Kursplan för matematik* [Syllabus for Mathematics]. www.skolverket.se/sb/d/2386/a/16138. 22 July 2010.

Swedish National Agency for Education. (2011a). *Del ur Lgr 11: kursplan i modersmål för grundskolan* [Part of Lgr 11: Syllabus for Mother Tongue for compulsory school]. www.skolverket.se/content/1/c6/02/21/84/Modersmal. pdf. 2 Feb 2011.

Swedish National Agency for Education. (2011b). *Om nationella prov* [About national tests]. www.skolverket.se/sb/d/2852. 2 Feb 2011.

Swedish National Agency for Education. (2011c). *Läroplan för grundskolan, förskoleklassen och fritidshemmet: del 1 och 2* [Curriculum for the compulsory school system, pre-school class and the leisure-time centre: Part 1 and 2]. www. skolverket.se/content/1/c6/02/38/94/Lgr11_kap1_2.pdf. 1 Apr 2011.

Swedish National Institute of Public Health. (2002). *Skolbarns hälsovanor 2001/2002* [School children's health habits 2001/2002]. Report number R 2003:50. Stockholm: Swedish National Institute of Public Health.

Swedish Work Environment Authority (2013). Bullrig arbetsmiljö sänker prestationsnivån i skolan [Noisy work environment lowers the level of performance in school]. https://www.av.se/. Accessed 23 Apr 2020.

The Bible. (2006). Bibelkommissionens översättning [Bible Commission translation]. Örebro, Sweden: Publishing House Libris.

The National Encyclopedia. (2010a). Tystnad, [Silence], www.ne.se.proxy.lib.ltu. se/sve/tystnad?i_h_word=tystnad. 26 June 2010.

The National Encyclopedia. (2010b). Ljud, [Sound]. www.ne.se.proxy.lib.ltu.se/ lang/ljud/243395. 24 July 2010.

The National Encyclopedia. (2010c). Retreat. www.ne.se.proxy.lib.ltu.se/lang/ retreat. 23 July 2010.

The National Encyclopedia. (2010d). Lyssna, [Listen]. www.ne.se.proxy.lib.ltu.se/sve/lyssna?i_h_word=lyssna. 22 July 2010.

United Nations Convention on the Rights of the Child. (1989). UN General Assembly Document A/RES/44/25.

van Manen, M. (1990). In State University of New York Press (Ed.), *Researching lived experience. Human science for an action sensitive pedagogy.* London.

Verschueren, J. (1985). *What do people say they do with words: Prolegomena to an empirical-conceptual approach to linguistic action.* Norwood, NJ: Ablex.

Von Wright, M. (2012). Silence in the asymmetry of educational relations. In H. Fiskå Hägg & A. Kristiansen (Eds.), *Attending to silence. Educators and philosophers on the art of listening.* Kristiansand, Norway: Portal Academic.

Vygotsky, L. S. (1978). *Mind in society. The development of higher psychological process.* Cambridge, MA: Harvard University Press.

Walter, G. (1995). *Bonniers synonymordbok* [Bonniers thesaurus]. Stockholm: Publishing House Bonnier Alba AB.

Waterhouse, A. L. (2013). *I materialenes verden; perspektiver og praksiser i barnehagens kunstneriske virksomhet* [In the world of materials; perspectives and practices in kindergarten artistic activities]. Bergen, Norway: Fagbokforlaget.

Weil, S. (2005). *Attention and will. An anthology.* London: Penguin Books.

Wiersma, W., & Jurs, S. G. (2005). *Research methods in education: An introduction.* Boston: Pearson.

Williams, J., & Currie, C. (2000). Self-esteem and physical development in early adolescence: Pubertal timing and body image. *Journal of Early Adolescence, 20*(2), 129–149.

Wittgenstein, L. (1922). *Tractatus Logico-Philosophicus.* London: Kegan Paul Trench, Trubner & CO.

Index[1]

[1] Note: Page numbers followed by 'n' refer to notes.

© The Author(s) 2020
E. Alerby, *Silence within and beyond Pedagogical Settings*,
https://doi.org/10.1007/978-3-030-51060-2

101